D0425231

THEY KNEW

THEY KNEW

The Federal Government's Fifty-Year Role in Causing the Climate Crisis

JAMES GUSTAVE SPETH

An Our Children's Trust Book

The MIT Press
Cambridge, Massachusetts
London, England

This book was set in Adobe Garamond Pro by New Best-set Typesetters Ltd. Printed and bound in the United States of America.

Library of Congress Cataloging-in-Publication Data

Names: Speth, James Gustave, author.
Title: They knew : the federal government's fifty-year role in causing the climate crisis / James Gustave Speth.
Description: Cambridge, Massachusetts : The MIT Press, 2021. | Includes bibliographical references and index.
Identifiers: LCCN 2020050507 | ISBN 9780262542982 (paperback)
Subjects: LCSH: Environmental policy—United States—History—20th century. | Environmental policy—United States—History—21st century. | Climatic changes—Government policy—United States—History—20th century. | Climatic changes—Government policy—United States—History—21st century.
Classification: LCC GE180 .S638 2021 | DDC 363.738/74560973—dc23
LC record available at https://lccn.loc.gov/2020050507

10 9 8 7 6 5 4 3 2 1

For the 21 Youth Plaintiffs in *Juliana v. United States*
Nathan Baring
Vic Barrett
Zealand Bell
Jaime Butler
Levi Draheim
Jayden Foytlin
Tia Hatton
Kelsey Cascadia Rose Juliana
Sophie Kivlehan
Jacob Lebel
Alexander Loznak
Xiuhtezcatl Tonatiuh Martinez
Avery McRae
Kiran Oommen
Aji Piper
Hazel Van Ummersen
Sahara Valentine
Nick Venner
Isaac Vergun
Miko Vergun
Journey Zephier

Contents

Introduction

Julia Olson and Philip Gregory

Since 2018, we have been blessed to work with Gus Speth, who did a remarkable job in developing his expert report for our constitutional climate lawsuit, *Juliana v. United States*. As you will see, *Juliana* "is no ordinary lawsuit." This introduction describes the story behind *Juliana* and explains the invaluable historical importance of this book to the overall strategy of the litigation. We also portray how *Juliana* came to be filed.

The original expert report was prepared by Gus as an expert witness in the *Juliana* litigation. His report (and now this book) is the most compelling indictment yet written of the federal government's role in the climate crisis. His book achieves its impact, in part, by a detailed presentation of original source material drawing on extensive research of the record over the past fifty years, mainly, but not entirely, government documents stretching from Presidents Johnson through Trump. As explained in the Reader Access to Exhibits Section, the associated exhibits can be digitally accessed by those interested. His book updates the original expert report with new insights about the Trump administration, such as various aspects of the extremely aggressive agenda in favoring fossil fuel energy over energy production that would halt climate change,[1] and its unparalleled harm to children and their health. Yet while the Trump administration's record has been deplorable, this historical analysis points out that, in many respects, the Trump administration's record is not much worse on energy or exacerbating the climate crisis than some prior administrations, and generally continues forward the stamp of approval for fossil energy that every president before Trump has likewise backed. In urging its publication, we believe Gus's expert report deserves a wide audience, now more than ever, as a historical lesson and a warning: even with a new Democratic administration, it will take much more than undoing President Trump's anti-climate-action policies to redress the long-standing federal government support of fossil energy that has led to the climate crisis.

The climate issue continues to be front and center politically and in the courts. This book is a compelling telling of the fifty-year history that is the backdrop to today's unfolding disaster. As we await the day Gus will take the stand and swear to tell the truth of this report before a trial judge in Eugene, Oregon, we hope this book and international interest in *Juliana* will catch the wave of rising public concern and political momentum. Moreover, looking further down the road, possibly for decades, people will be asking how the climate tragedy happened. What did the federal government know and when did it know it? In the face of this information, what did the federal government do and what did it not do? This book addresses these questions. The public should have this material readily available. Among other things, it documents convincingly that the federal government knew enough in the 1970s and 1980s to begin addressing the climate issue in energy policy and elsewhere. Hopefully, after reading this book, you will understand why people such as Bill McKibben and Naomi Klein call *Juliana* "the most important lawsuit on the planet."

BACKGROUND ON OUR CHILDREN'S TRUST

This book was prepared in conjunction with Our Children's Trust (OCT). Julia Olson founded OCT in 2010 after becoming a mother. She quickly grew frustrated at the limitations of environmental law, and of her own public interest legal practice, in addressing the growing climate crisis. Her young children were born into an increasingly hotter planet and she was determined to use a rights-based legal approach to advocate more holistically and strategically for the system change needed to thwart the crisis—to give them, and all children, a shot at a safe and livable world.

Julia's bible in those early days of creating the legal strategy that would lead to *Juliana* was *Storms of My Grandchildren*, a book published in December 2009 by Dr. James Hansen, one of the world's most respected climate scientists. (Dr. Hansen's granddaughter, Sophie, is now one of the plaintiffs in *Juliana*, and Dr. Hansen serves as a plaintiff Guardian for Future Generations.) At the heart of any case about climate crisis is the science, and Julia devoured the literature. The legal theory would come next, inspired by the synchronous scholarship of Professor Mary Wood, author of *Nature's Trust*.

Thus, in 2010, OCT was formed. At that point, our average global atmospheric CO_2 level was already at 389 parts per million and swiftly rising well beyond the 350 parts per million threshold that would keep our planet's climate system from spiraling out of balance.[2] There was urgency to take bold action on behalf of children and future generations. With the guidance of revolutionary legal scholars like Mary Wood and leading climate scientists like Dr. Hansen, Julia, alongside a few other lawyers, such as Phil Gregory and Andrea Rodgers, leapt into action to launch Our Children's Trust's legal strategy.

By Mother's Day 2011, OCT had prepared to file climate lawsuits or regulatory legal actions in fifty states—as well as a federal case—all on behalf of young people.

As of 2020, there is still no other public interest law firm of this kind on our planet. Representing only youth, from many different communities and cultures. Suing only governments to protect fundamental human rights and equal protection of the law. Seeking only science-based systemic remedies. We at OCT live by these principles and we work together to do one thing: save our children's only planet from government-sanctioned climate destruction.

OCT is a nonprofit organization elevating the voice of youth around the globe to secure the legal right to a healthy atmosphere and stable climate system.[3] OCT supports youth climate advocates in legal actions against federal and state governments, as well as internationally, advancing constitutional and public trust rights to a safe climate. OCT seeks court-ordered government implementation of systemic science-based Climate Recovery Plans that will reduce atmospheric CO_2 concentrations to below

350 parts per million before the year 2100, in accordance with expert scientific prescriptions for climate stabilization.

With the invaluable support of many people around the globe, OCT has accomplished so much since it was founded in 2010. OCT has been involved with hundreds of young people and youth organizations bringing landmark climate litigation, and thousands more in other legal actions. In 2020, OCT has active legal actions moving forward in seven states—Alaska, Washington, Oregon, Montana, Colorado, North Carolina, and Florida—with actions in more states in development. OCT also is involved in proceedings in several state administrative agencies. Every case has built upon the last case, trailblazing the way to critical legal victories of undisputed scientific evidence and new legal precedent in our courts. Then there is OCT's global work, supporting and inspiring youth climate cases in more than twenty other countries, including India, Pakistan, Uganda, Canada, Colombia, and Mexico. Underlying these actions are state and global domestic court rulings from related cases demanding government climate action, many brought by youth with the support of OCT. OCT is also actively working with youth and lawyers in multiple additional state and global domestic jurisdictions, to coordinate filing of additional parallel constitutional and public trust legal actions against those governments in the near future, each seeking court-mandated science-based government action to mitigate the climate crisis.

WHY WE FILED *JULIANA*

The uncontradicted evidence prior to trial established that the United States is responsible for one-quarter of the accumulated CO_2 in the atmosphere.[4] For more than a hundred years, scientists have understood that burning fossil fuels caused CO_2 emissions and increasing atmospheric CO_2 levels caused climate change.[5] As Gus's book makes abundantly clear, for at least fifty years, the federal government, all the way up to the White House, has understood the climate science and issued reports on the catastrophic dangers of continuing to burn fossil fuels.[6] For decades, in spite of this knowledge, the defendants, under both Republican and Democratic leadership, have knowingly promoted and controlled a national fossil fuel energy system when available alternatives existed.[7] One of the plaintiffs' experts in *Juliana*, Nobel laureate Dr. Joseph Stiglitz, confirms that "the current national energy system, in which approximately 80 percent of energy comes from fossil fuels, is a direct result of decisions and actions taken by Defendants."[8] In his expert opinion, Dr. Stiglitz avers:

> The fact that the U.S. national energy system is so predominately fossil fuel-based is not an inevitable consequence of history. The current level of dependence of our national energy system on fossil fuels is a result of intentional actions taken by Defendants over many years. These actions, cumulatively, promote the use of fossil fuels, contribute to dangerous levels of CO_2 emissions, and are causing climate change.[9]

It is important to understand that, in *Juliana*, the plaintiffs challenge the defendants' fossil fuel energy system on a systemic basis. The plaintiffs do not individually challenge isolated acts within that system. Nor do they challenge the defendants' failure to ban fossil fuels or only claim the defendants "should have done more." Rather, the plaintiffs challenge the defendants' *affirmative* actions in promoting and perpetuating the use of fossil fuels as the centerpiece of the nation's energy system, in spite of knowing the catastrophic consequences. The nation's dependence on fossil fuels could not have occurred but for the defendants' unconstitutional conduct.

An example of systemic litigation similar to *Juliana* is in the prison reform litigation in California that led to the US Supreme Court's decision in *Brown v. Plata*.[10] There, California prisoners argued that their rights under the Eighth and Fourteenth Amendments were being violated. The California prisoner population was vastly exceeding the capacity of the prisons, at 200 percent capacity, jeopardizing the safety and health of those incarcerated. After hearing from experts, including scientists, psychologists, physicians, and engineers, and looking at the capacity of the prisons, the Supreme Court set a standard for constitutional compliance. Based on the expert testimony, the courts settled on a maximum of 137.5 percent capacity for California's prisons to rectify the constitutional infringement of prisoners' rights.

Similarly, in the context of *Juliana*, the court needs to set a standard for levels of atmospheric CO_2 to which our nation's energy policies should be directed to return in order to protect young people and future generations. What the judiciary can do is identical to what the court did in *Brown v. Plata*: hear testimony from experts about what those safe levels of CO_2 should be and then require development and implementation of a national climate recovery plan based on that evidence. The uncontradicted expert testimony in *Juliana* presents extensive physical science on what different levels of CO_2 lead to in terms of temperature increase and sea-level rise. Our experts testified *without contradiction* that our nation initially needs to be on a path to return to 350 parts per million of atmospheric CO_2 or below by 2100; however, our foremost ice sheet and sea-level rise experts also testified that, in subsequent centuries, our nation really needs to head back toward preindustrial levels of CO_2 in the atmosphere: 280 parts per million.

In other countries, courts have set standards and then required the government to come into compliance with emission levels that are on track to return to safe CO_2 levels. For example, in 2019, the Netherlands Supreme Court in *Urgenda Foundation v. Netherlands*[11] held the government had an obligation to do more to avert the imminent danger posed by climate change, and set tighter standards than its international treaty contributions, despite the Netherlands being only minimally responsible for the greater increase of greenhouse gas (GHG) emissions globally.

One of the fundamental legal arguments in *Juliana* is courts sit in equity and they can, based on the evidence, creatively develop remedies to redress the constitutional injury, while respecting the role of the political branches. Such orders are issued by courts all the time. How the executive and legislative branches decide to come into constitutional compliance is an issue separate from the court requiring the defendants to decarbonize in order to protect fundamental rights.

While it is possible that Congress could step in with a legislative solution to swiftly decarbonize our energy system if it wished, as it intervened with the Civil Rights Act and the Voting Rights Act to rectify constitutional infringements during the Civil Rights movement, the executive branch also has broad authority under existing law to implement a court-ordered remedy. The ability of the executive branch to control our nation's energy system is bolstered by findings of the District Court Judge, Ann Aiken, in *Juliana*, when she denied the defendants' motions for judgment on the pleadings and summary judgment in October 2018. In terms of control, Judge Aiken found (1) the plaintiffs "proffered uncontradicted evidence showing that the government has historically known about the dangers of greenhouse gases but has continued to take steps promoting a fossil fuel based energy system, thus increasing greenhouse gas emissions," and (2) "the pattern of federally authorized emissions challenged by plaintiffs in this case do make up a significant portion of global emissions." Because of the defendants' fossil fuel–based energy system, CO_2 emissions during 1850–2012 from the United States constituted more than one-quarter of cumulative global CO_2 emissions. Leading up to trial, the federal government never disputed that a national energy system exists and presented no evidence to refute the defendants' control over the fossil fuel–based energy system and the GHG emissions that result therefrom.

Today, when it is technically and economically feasible to transition swiftly away from fossil energy, and when the climate system is in a dangerous state of emergency, the defendants are recklessly increasing fossil fuel development.[12] "The United States is expanding oil and gas extraction on a scale at least four times faster and greater than any other nation and is currently on track to account for 60% of global growth in oil and gas production."[13] As part of the defendants' fossil fuel energy system and strategy for fossil fuel dominance, there are presently (in 2019) close to a hundred new fossil fuel infrastructure projects poised for federal permits, including pipelines, export facilities, and coal and liquefied natural gas terminals.[14] Such conduct threatens our national security.[15] "The economic impacts of these actions are deleterious to Youth Plaintiffs and the Nation as a whole. Defendants' actions promoting a fossil fuel–based energy system are serving to undermine the legitimate government interests of national security and economic prosperity that they purport to advance."[16] These enormous economic burdens and health costs will be borne by these plaintiffs, other children, and generations to come.[17]

As we write this Introduction in 2020, wildfires rage along the West Coast and hurricane season peaks on the East Coast, threatening many of our *Juliana* plaintiffs' health, homes, and security. Many of OCT's staff members as well as our youth plaintiffs are currently suffering from the most hazardous air quality on the planet as toxic smoke and ash fill the skies and seep into our homes, and as encroaching fires force evacuations. These climate fires have harmed our own health and destroyed the farmland and livelihood of Phil's son. Simultaneously, our friends on the East Coast have experienced a record-breaking hurricane season that has been twice as active as usual, and many are currently facing flash flooding due to heavy storms. Extreme weather events and natural disasters will only increase as the climate crisis worsens. Communities across the United States will increasingly experience these cataclysmic extreme weather events and natural disasters. That's why we are committed to representing young people from coast to coast as they fight to protect their legal right to a stable climate.

Since she was four years old, plaintiff Jaime has been working to protect the Earth. Jaime is of the Tangle People Clan, born for the Bitter Water Clan; her maternal grandfathers are the Red House Clan, and her paternal grandfathers are the Towering House Clan. She grew up in Cameron, Arizona, on the Navajo Nation Reservation. In 2011, Jaime and her mother had to move from Cameron to Flagstaff because of water scarcity. Jaime and her extended family on the reservation remember times when there was enough water on the reservation for agriculture and farm animals, but now the springs they once depended on year-round are drying up. Jaime also sees firsthand the cultural and spiritual impacts of climate change, as participating in sacred Navajo ceremonies on the reservation is an important part of her life, and climate impacts are starting to harm the ability for Jamie and her tribe to participate in their traditional ceremonies.

The devastating COVID-19 public health pandemic and a renewed national focus on systemic racial injustice further highlight the urgency of addressing climate crisis and how it intersects with environmental injustice—because children and future generations from all communities are affected by the devastation, especially vulnerable communities hardest hit by the climate crisis. Black, Brown, and Indigenous youth bear the largest burden of government-waged harm.

The plaintiffs, as youth, are especially vulnerable to the impacts of climate change.[18] Plaintiff Journey's personal well-being depends upon the coral reefs in Hawaii that are dying at accelerating rates.[19] The extreme storm events that have already harmed individual plaintiffs, including Levi, Journey, and Jayden, are becoming increasingly frequent and destructive and will get worse without immediate action to reduce GHG emissions.[20] Levi has been forced to evacuate his home on a barrier island off the coast of Florida because of hurricanes and flooding, which are driven by increased ocean heat content and the resulting high temperatures of the sea's surface.[21] Fleeing from his home, having his school permanently closed after Hurricane Irma, and witnessing climate-change-induced environmental devastation have caused Levi to legitimately fear for his personal safety and security.

The plaintiffs also are bearing the health burdens of climate change. An expert on the health effects of climate change, Dr. Jerome Paulson finds the defendants' actions "truly shocking" in light of the "undisputed health risks to children." "By continuing to promote fossil fuels, the federal government is knowingly putting these children in an increasingly risky situation when it comes to their health."[22] Some of the plaintiffs are "at risk of irreparable harm from having decreased lung function as a result of growing up in environments with more air pollution." Plaintiffs like Nick who have asthma are already harmed by pollution from fossil fuels, increased prevalence of wildfire smoke, and exacerbated ozone conditions due to climate change.[23] The more fossil fuels are burned, the worse Nick's health will be. "Without immediate and significant actions to reduce greenhouse gas emissions by Defendants, global temperatures will continue to increase and exacerbate [wildfire] conditions. The magnitude of wildfire that destroyed Paradise, California, is a harbinger of destruction to come in the West."[24] Dr. Paulson warns: "For Sahara, Jacob, Alex, Isaac, Aji, Nicholas, and other Plaintiffs exposed to smoke from wildfire, I expect, consistent with the literature, that their increased exposure to smoke will exacerbate existing health issues, such as asthma, and may cause new acute and chronic respiratory illnesses. By continuing to promote fossil fuels, the federal government is knowingly putting these children in an increasingly risky situation when it comes to their health."[25]

The plaintiffs are also being profoundly psychologically harmed by the defendants. As noted by one of our experts, Dr. Lise Van Susteren, a psychiatrist with a special interest in the psychological effects of climate change: "Climate change is causing devastating physical impacts—injuries, illnesses, and deaths. But for the magnitude of its impacts, the potential insinuation into every aspect of our lives, the relentlessness of its nature and debilitating effects, it is the emotional toll of climate change that is even more catastrophic, especially for our children." [26] The climate crisis has the capacity to destroy children psychologically.[27] Plaintiffs Aji, Nick, Levi, and Journey all attest to intense impacts on their mental and emotional well-being. Sleeplessness, nightmares, anxiety, anger, depression, fear, and deep feelings of betrayal by their government are part of the psychological makeup of these young people.[28]

Trauma from climate change and institutional betrayal can alter hormone levels, brain development, cognitive functioning, reproductive success, and even children's DNA.[29] According to Dr. Lise Van Susteren, "The only way to relieve at least part of the psychological harm Plaintiffs are experiencing from the federal government's institutional betrayal is for the government to stop endangering Plaintiffs."[30]

THE IMPORTANCE OF THE SPETH EXPERT REPORT

The Due Process Clause of the US Constitution provides no person shall be deprived "of life, liberty, or property, without due process of law." In most cases, this means the government must provide adequate safeguards before it can restrict one of these rights.[31] But some rights—referred to by the courts as fundamental rights—are so important that no amount of procedure alone will do. Rather, the government can infringe upon these rights only if its imposition "is narrowly tailored to serve a compelling state interest," a doctrine referred to as substantive due process.[32] The right to life and safety through personal security is such a fundamental interest, and therefore is protected by the substantive portion of the Due Process Clause.[33] At a fundamental level, one of the rights at issue in *Juliana* is the right to personal security, perhaps the pinnacle of an individual's interests in life and liberty.[34]

That said, the US Constitution concerns the actions of the government, not private citizens. Thus, while the government cannot infringe upon a fundamental right without a compelling state interest, the government generally is not obligated to protect those rights against harm from private actors. That is the central holding of the US Supreme Court in *DeShaney v. Winnebago County Department of Social Services*.[35]

In *DeShaney*, the Supreme Court held: "As a general matter, . . . a State's failure to protect an individual against private violence simply does not constitute a violation of the Due Process Clause."[36] The Supreme Court noted two exceptions to this rule, saying: "In certain limited circumstances the Constitution imposes upon the State affirmative duties of care and protection with respect to particular individuals."[37] First, "when the State takes a person into its custody and holds him there against his will, the Constitution imposes upon it a corresponding duty to assume some responsibility for his safety and general well-being."[38] This is the custody exception to the general rule in *DeShaney*. Second, the Supreme Court noted: "While the State may have been aware of the dangers that Joshua [DeShaney] faced in the free world, it played no part in their creation, nor did it do anything to render him any more vulnerable to them."[39] The circuit courts, including the Ninth Circuit, have interpreted this statement as providing another exception to *DeShaney* where the state acts to create or increase the danger of private harm: the "state-created danger" doctrine.[40]

A state-created danger claim under the Due Process Clause arises where (1) "the state affirmatively places the plaintiff in danger" and (2) the state "act[s] with 'deliberate indifference' to a 'known or obvious danger.'"[41] To establish "deliberate indifference," a plaintiff must show (1) the defendants' actual knowledge of or willful blindness to (2) an unusually serious risk of harm and (3) the defendants either failed to take obvious steps to address the risk or exposed a claimant to the risk.[42] Deliberate indifference is shown by evidence that a governmental actor "'disregarded a known or obvious consequence of his action.'"[43] A plaintiff must establish "the state engaged in 'affirmative conduct' that placed him or her in danger."[44] Affirmative conduct is conduct that creates, exposes, or increases a risk of harm that a plaintiff would not have faced to the same degree absent such conduct.[45]

In reviewing the evidence on affirmative conduct, courts address whether the government's generalized "affirmative acts created a danger" that the plaintiff "otherwise would not have faced." As the Ninth Circuit wrote in *Hernandez v. City of San Jose*, an action brought under 42 U.S.C. §1983 against police officers by attendees of a political rally for Donald Trump who were attacked by anti-Trump protesters as they attempted to leave the rally: "Being attacked by anti-Trump protesters was only a possibility when the Attendees arrived at the Rally. The Officers greatly increased that risk of violence when they shepherded and directed the Attendees towards the unruly mob waiting outside the Convention Center." The Ninth Circuit then went on to state that, to show deliberate indifference, "the Attendees must [demonstrate] the Officers 'recognize[d] [an] unreasonable risk and actually intend[ed] to expose [the Attendees] to such risks without regard to the consequences to [the Attendees].'"[46] "In other words, the [Officers] [must] [have] 'known that something [was] going to happen but ignor[ed] the risk and expose[d] [the Attendees] to it [anyway].'"[47]

As this book ably demonstrates, the federal government's long-standing knowledge of the profound risks of climate destabilization from continued fossil fuel use, and the resulting harms to the plaintiffs, are extensively recorded in federal government documents spanning decades. As Dr. Hansen declared in *Juliana*: "The great danger for young people, is that they are being handed a situation that is out of their control, a situation made more egregious due to the fact that the Defendants have a complete understanding of precisely how dangerous the situation is that they are handing down to these Plaintiffs."[48]

Before trial in *Juliana*, the federal government claimed "Plaintiffs identify no specific government actions—much less government actors—that put them in 'obvious, immediate, and particularized danger.'"[49] Again, that assertion is belied by Gus's book. As this book shows, the defendants' affirmative conduct with respect to fossil fuels is resulting in greater CO_2

emissions levels and concentrations than would occur absent such conduct. Other uncontradicted expert testimony in *Juliana* establishes that excess CO_2 emissions resulting from the defendants' conduct continue to destabilize the climate, resulting in mounting injuries to plaintiffs.[50] "Cumulative emissions by the United States substantially exceed those of any other nation. Thus, the United States is, by far, more responsible than any other nation for the associated increase of global temperature."[51] The plaintiffs have shown the defendants substantially caused and contributed to dangerous climate destabilization and the already-occurring and imminently threatened harms faced by the plaintiffs.[52]

The plaintiffs do not contend the defendants are the sole contributors to climate change, nor do they need to be for the plaintiffs to prevail. In *Juliana*, the defendants admit they affirmatively "permit, authorize, and subsidize fossil fuel extraction, development, consumption, and exportation"; that emissions "from such activities have increased the atmospheric concentration of CO_2"; that the United States is responsible for more than a quarter of global historic cumulative CO_2 emissions; and that "current and projected atmospheric concentrations of six well-mixed GHGs, including CO_2, threaten the public health and welfare of current and future generations, and this threat will mount over time as GHGs continue to accumulate in the atmosphere and result in ever greater rates of climate change."[53]

As this book's section on the Trump administration shows, the defendants continue to affirmatively double down on the use of fossil fuels. The United States is among the world's largest producers of fossil fuels, and is the world's single largest producer of both oil and gas. A staggering amount of GHG emissions is caused by the defendants' leasing of federal public lands for fossil fuel extraction and production. From 2008 through 2017, US petroleum and natural gas production increased by nearly 60 percent. Since 2017, the defendants have opened vast areas of federal lands and waters for fossil fuel exploration and production. Presently, the defendants have "plans to allow new offshore oil and gas drilling in virtually all (98%) of U.S. coastal waters during 2019–2024."[54] The defendants are also poised to lease even more federal public lands for fossil fuel extraction and permit new oil and gas pipelines, liquefied natural gas and coal terminals, and deepwater port oil export facilities as part of the national fossil fuel energy system.

The defendants' affirmative conduct thereby placed the plaintiffs "in a situation more dangerous than the one" they would otherwise face.[55] Again, as Dr. Hansen observed, "Plaintiffs are already being harmed by Defendants' conduct, past and present, in causing substantial amounts of GHG emissions, but the harm continues to worsen with increasing amounts of fossil fuel development and promotion of fossil fuel energy."[56]

With respect to the third component of deliberate indifference, this book establishes the defendants have refused for decades to take obvious steps to address the profound harms and unprecedented dangers, ignoring technologically and economically feasible alternative energy pathways.[57]

The documents underpinning this book show the defendants' present conduct recklessly disregards the substantial risk of harm to the plaintiffs and the nation.[58] The uncontroverted evidence shows the defendants were aware of the dangers to the plaintiffs. In fact, the defendants knew that continuing to accelerate fossil fuel emissions posed a threat to the plaintiffs. Yet, like the police officers in *Hernandez*, they continued to "direct[] [the Attendees] into the mob."[59] Gus's book is a cornerstone of the evidence, establishing a due process violation pursuant to the state-created danger doctrine and demonstrating the defendants "act[ed] with deliberate indifference to a known [and] obvious danger."[60] That government conduct has placed us all in an existential emergency, one our children implore us to fix.

CONCLUSION

We would like to tell you a story that has inspired us. It is from the autobiography of the Rev. Dr. Pauli Murray.[61] In 1944, she was attending the all-African American law school at Howard University. She was the only woman in her class; she was first in her class in terms of grades. This was during the period when the work of the National Association for the Advancement of Colored People (NAACP) on racial discrimination and segregation was focused on challenging the inequality in public facilities, in other words, only the "equal" side of *Plessy v. Ferguson*'s "separate but equal" standard,[62] and trying to get equal facilities and funding for African American communities and children. In discussing the NAACP's work one day in her law school class, Pauli Murray argued to her professor and the class that a case should be brought to challenge "separate" as unconstitutional. Everyone laughed at her. She articulated her argument and she wrote a paper on it. But she also bet her law professor ten dollars that within twenty-five years, *Plessy* would be overturned. Nobody in the class believed her at the time.

That same law professor went on to serve as co-counsel in *Brown v. Board of Education*[63] with Thurgood Marshall and used Pauli's work from her law review paper to inform the claims they brought in *Brown*. *Brown*, and *Mendez v. Westminster School District of Orange County*[64] before *Brown*, which desegregated California schools, emanated from children saying, "We want to go to that school." These cases were not just attorneys and parents advocating for children. It was youth rising up, saying "enough," in ways similar to what young people are doing today in the face of injustice. The youth then and the youth today are saying "enough is enough, we don't want this system anymore—it does not serve us." There are amazing stories about some of the young plaintiffs in *Brown*, also at the forefront of mobilizing for their rights. There are also incredible stories of the courageous judges who were responsible for enforcing *Brown*, in what was a social revolution linked to political and economic transformation. Despite a period of turmoil, disruption, and instances of violent upheaval, the courts saw change must occur. The rule of law prevailed.[65]

Like the youth plaintiffs in *Juliana*, from a young age Pauli Murray also thought outside the box and was a pioneer for justice. She inspired legal thinking that informed Ruth Bader Ginsburg's work seeking equal rights for women, in a nation dominated by the political power of men. Pauli Murray's story is one that should never be lost.

Pauli Murray was also a poet. In *Dark Testament and Other Poems*, she wrote:

> Hope is a crushed stalk
> Between clenched fingers.
> Hope is a bird's wing

Broken by a stone.
Hope is a word in a tuneless ditty—
A word whispered with the wind,
A dream of forty acres and a mule,
A cabin of one's own and a moment to rest,
A name and place for one's children
And children's children at last . . .
Hope is a song in a weary throat.
Give me a song of hope
And a world where I can sing it.
Give me a song of faith
And a people to believe in it.
Give me a song of kindliness
And a country where I can live it.
Give me a song of hope and love
And a brown girl's heart to hear it.[66]

Gus Speth, too, is a poet.[67] You will see from his book that he has a gift of bringing this recent, woeful history of our federal government vividly to life, telling us much that we had not known about our national energy system. As Gus frames it through his poetic lens: "Focusing on this history, no matter how extensive its scope, we have to see this period of almost half a century as an instant in time, for that is indeed what it is in Earth's time and in the lives of today's children and all those who must live with its consequences. And in this snapshot of decades, we find a federal government planning for, guiding, supporting, and encouraging massive fossil fuel use despite tragic consequences easily foreseen and avoided."[68] We believe, after reviewing this book, you too will conclude that by causing and contributing to dangerous climate change through affirmative aggregate acts and policies with respect to fossil fuels and the national energy system, the federal government has violated these youth plaintiffs' substantive due process rights to life, liberty, and property, as well as previously recognized unenumerated substantive due process rights, including rights to bodily integrity and personal security and the right to family autonomy, including the right to safely raise a family, to keep one's family together, and to learn and transmit one's spiritual and cultural heritage and traditions, in each case guaranteed by the Fifth Amendment of the US Constitution. To quote Gus, the evidence is clear: By knowingly engaging in historic and ongoing conduct that has placed the plaintiffs in a position of danger, the defendants' aggregate acts and policies with respect to fossil fuels and the national energy system have knowingly exposed these plaintiffs to danger, with deliberate indifference to their safety, instead making the conscious shocking decision to "promote fossil fuels and thus to cause irreversible climate danger, a pattern that can only reflect a deliberate indifference to the severe impacts that will follow—impacts to be endured predominantly by youth plaintiffs and future generations."[69]

Our constitutional climate work in *Juliana* will help set systemic constitutional standards and a decarbonization mandate that will have a profound effect on issues of environmental and racial justice. It is through experts such as Gus Speth, all working *pro bono*, that our federal judiciary will have the evidence both to determine whether the rights of our youth plaintiffs have been infringed and to set a standard for righting the historic wrongs.[70] When courts find that there is a right, such as a right to a climate capable of sustaining human life, and that the right has been infringed, judges must set a standard for government coming into compliance with our Constitution and upholding that right. Often, scientific and historical evidence, such as is presented by our experts, informs that right and the realization of our liberties.

Natural resources do not "belong" to any generation; they are to be preserved, in trust, for all generations. Throughout his career, Gus Speth has had a significant role in protecting our Earth's atmosphere and natural systems for present and future generations. We hope, after reading his expert report, you will understand why the climate crisis, as a moral and legal obligation, has required us to turn to the courts to hold our federal government legally accountable for the harms caused by its prior and current actions. We will continue to address the climate crisis and give youth a strong voice in the climate debate. We hope this book will inspire you to join us.

Author's Note

When Julia Olson, co-lead counsel in *Juliana v. United States* and the executive director of Our Children's Trust, asked me to write this expert report for the *Juliana* litigation, I was both surprised because I did not see it coming and delighted because of the opportunity she provided not only to participate in a great cause but also to tell the story of the failure of leadership that has led us to the climate emergency we now face. I will always be grateful to her and to her co-counsel Phil Gregory for that opportunity and for the amazing support they have provided throughout.

I prepared the report mostly in 2018. What you have before you is that report with only modest updates. Most importantly, the chapter on the Trump years is largely new. As for the text leading up to these last few years, I have wanted to keep it as originally submitted in the litigation in 2018 as my expert report. I have made some exceptions to this goal, mostly to incorporate new material brought to my attention.

Julia and Phil are not the only ones who were a big help. Ben Jervey, Allison Kole, Nate Bellinger, Daniel Noonan, Eli Brown, Susan Carey, Andrea Rodgers, Matt McRae, Devin Kesner, Emily Clark, Nan Leuschel, and Lou Helmuth were as well. I'm sure there were countless hours spent by law clerks and volunteer lawyers, doing such tasks as helping to curate the exhibits. I did not meet everyone involved in providing support, but I know they were there and I thank each and every one. And as with all my books, my partner for fifty-five years, Cameron Speth, was once again indispensable.

This section would not be complete without expression of our great appreciation and admiration for the editorial team at the MIT Press. Beth Clevenger, Anthony Zannino, and Judith Feldmann in particular gave talented and most welcomed help throughout.

I am thinking now of the effort back in 2018 to produce a report covering fifty years and containing 464 footnotes and 292 exhibits. For their parts in this effort, those mentioned here have earned my admiration. Also noteworthy is the enormous contribution in documentary research done by the remarkable Our Children's Trust staff and volunteers. As a nonprofit organization providing free legal services to young people, I am well aware they could not do what they do without countless *pro bono* hours (mine included), as well as the visionary individuals and foundations that provide funding to support their work. My thanks to them all.

That addresses the perspiration. The inspiration was never hard to find. It was always the twenty-one feisty, articulate young people who stepped forward to sue the federal government and demand their rights. Young people are now at the vanguard of the climate justice struggle. Starting in 2015, the *Juliana* plaintiffs have shown the way.

James Gustave Speth
Strafford, Vermont
September 2020

THE SPETH EXPERT REPORT FOR *JULIANA V. UNITED STATES*

Originally submitted September 28, 2018; updated September 28, 2020

OVERVIEW

I have been retained *pro bono* by plaintiffs to provide expert testimony regarding the historical knowledge of the US federal government (including defendants) of climate change, climate science, and alternative pathways to power the nation's energy system other than fossil fuels. I will also testify about the decisions made by the US federal government to devise and pursue energy policies and, in particular, to maintain a fossil fuel based energy system.[1]

By the end of the Carter administration in January 1981, almost four decades ago, it was already very clear that:

- defendants knew the basic science of climate change and knew that the continued burning of high levels of fossil fuels would lead to climate danger; and
- defendants knew of pathways recommended by experts within government and others to transition away from fossil fuels, including through conservation, efficiency, and solar and other renewables.

Notwithstanding this, defendants continued from the Carter years to the present to plan for, support, invest in, permit, and otherwise foster a national fossil–fuel-based energy system.

In the first part of this report, through documentary evidence, I will present some important background information that preceded the Carter administration in order to establish what the federal government knew about climate change when President Carter took office on January 20,

1977. I will then present the documentary evidence for my expert conclusions stated above as it existed at the end of the Carter administration, and I will supplement this evidence drawing on my personal participation and observation. I will take up each of the above points in turn. I used my contextual knowledge to present historical government evidence of the individual and institutional actors involved and the historical context of the events in question.

In the second part of this report, I will describe what transpired after President Carter left office with respect to the conclusions just noted: (1) that the federal government had the basic information about climate science and the link between fossil fuels and climate change, (2) that government also had abundant recommendations for reducing fossil fuel use as part of the national energy system, and (3) that notwithstanding those reasonable and available alternatives, the federal government continued to foster a fossil fuel based energy system. I will take up these three matters in the context of each administration following Carter, as with Carter but in less depth than the Carter presentation. Because the science of climate change and the in-depth understanding of the danger of fossil fuels only increased after the Carter administration, and are the subjects of other expert reports in this matter, I shall not attempt an exhaustive presentation.

One conclusion of this review stands out. The three-part pattern just noted continued through subsequent administrations and through the various Congresses: knowledge of the climate science and the dangers of fossil fuel burning, knowledge of alternatives to fossil fuels, and continued full-throttle support for development and use of fossil fuels.

For the year 1976, the year President Carter was elected, the United States relied on fossil fuels for 91 percent of primary energy consumption. In 2019, three years into President Trump's term, the United States was still overwhelmingly dependent on fossil fuels—80 percent. During this forty-three-year period, the seeds planted during the Carter administration regarding efficiency and renewable energy could have yielded a smooth transition toward an outstanding US climate performance and global leadership in climate action. Instead, those years saw only negligible action to actually reduce US fossil emissions and only modest actions to promote alternatives, with the result that US CO_2 emissions from energy consumption have gone up, not down, climbing by about 16 percent from 1975 to 2019.

Defendants' actions on the national energy system over the past several decades are, in my view, the greatest dereliction of civic responsibility in the history of the Republic. And it is worse today than ever. This shocking historical conduct, government malfeasance on a grand scale, has left current and future generations enormously vulnerable to substantial danger.

1 CLIMATE SCIENCE KNOWLEDGE THE CARTER ADMINISTRATION INHERITED

The 1930s through the early 1970s were a period of growing concern for scientists studying climate change. This period provides a prelude to the flowering forth of the climate change issue during the Carter administration.

Particularly by the 1960s, with the advance of computer modeling, scientists were able to verify the changes that were happening and project what could happen in the future as the atmospheric CO_2 concentration increased. This period was marked by the work of scientists such as Guy S. Callendar, Roger Revelle, Gilbert Plass, and Charles D. Keeling. By the late 1950s and early 1960s, these scientists started sounding alarm bells. With federally funded monitoring of the growing atmospheric CO_2 concentration, it was then possible to produce detailed charts showing the atmospheric CO_2 concentration. Early on, Keeling showed the seasonal variation in the concentration of atmospheric CO_2 in the Northern Hemisphere and that the CO_2 concentration in the late 1950s was around 315 ppm.[1] As the data were collected and charted year after year, the now famous "Keeling Curve" would become apparent, showing conclusively that the CO_2 concentration was on the rise (see figure 1.1).

In 1955, the United States established the first major climate modeling center, the Geophysical Fluid Dynamic Laboratory (GFDL) of the National Oceanic and Atmospheric Administration (NOAA), now located at Princeton University (but still a NOAA lab). In 1960, another center was established by the National Science Foundation: the National Center for Atmospheric Research (NCAR) in Boulder, Colorado. Besides

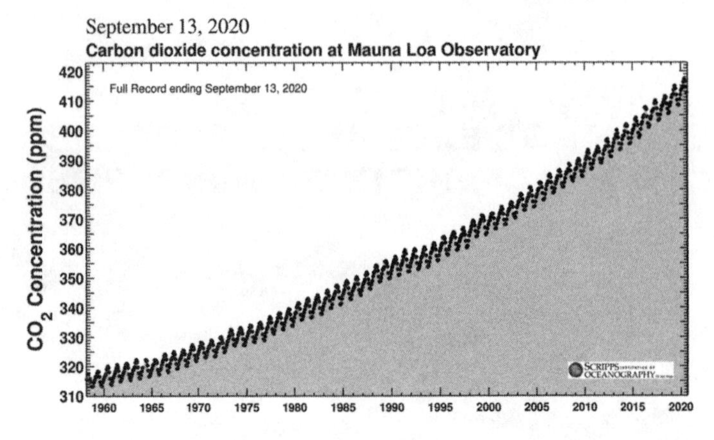

Figure 1.1

Carbon dioxide concentration at Mauna Loa Observatory, showing the Keeling curve. *Source*: Scripps Institution of Oceanography, "The Keeling Curve," https://scripps.ucsd.edu/programs/keelingcurve.

these government laboratories, UCLA established a modeling laboratory, as did the RAND Corporation, focused on possible military applications of climate modification. The RAND center was ultimately funded by the Defense Advanced Research Projects Agency (DARPA), which had been created in 1958 in response to the Soviet launching of Sputnik, with the mission of keeping US military technology ahead of any potential enemy. These centers were focused on better understanding and developing atmosphere–ocean climate models and the response of the climate system to an increased concentration of CO_2. During this time, the Department of Defense, through the Office of Naval Research, began to fund and conduct research pertaining to climate change.[2]

The year 1965 was a pivotal year for the federal government recognizing climate change, with key reports issued late that year. In November 1965, the White House issued the report of the Environmental Pollution Panel of the President's Science Advisory Committee, "Restoring the Quality of Our Environment," accompanied by Appendix Y4, the work of Roger Revelle et al. entitled "Atmospheric Carbon Dioxide."[3] The 1965 White House report found that increasing CO_2 levels would be deleterious to humans by the year 2000, would cause ocean warming, sea-level rise, and other adverse impacts, and that burning all fossil fuels would cause a 200 percent increase in the atmospheric CO_2 concentration. The White House created a task force and ordered responses to the issues raised. By May 1967, the Office of Science and Technology (OST) (the predecessor to today's Office of Science and Technology Policy (OSTP)) produced a report of agency responses to the 1965 White House report. The Department of Commerce reported that it is "studying the CO_2 content of the atmosphere" and "monitoring stratospheric temperature," with plans to augment its "CO_2 monitoring and evaluation program" during FY1967.[4]

Also at the end of 1965, on December 20, the National Science Foundation published the "Weather and Climate Modification: Report of the Special Commission on Weather Modification."[5] The "Weather and Climate Modification" report estimated that the CO_2 concentration in the atmosphere had increased 10–15 percent in the twentieth century, causing significant changes in Earth's heat balance.[6] The report states that "the implications of this upon tropospheric stability cannot be ignored" and that there is a need for continuous monitoring of CO_2 content and of simulation of CO_2 effects "using the most sophisticated atmospheric models and numerical computers available" to assess the consequences.[7]

In a 1966 National Academy of Sciences (NAS) report on "Weather and Climate Modification Problems and Prospects, Vol. I, Summary and Recommendations," the NAS generally agreed that warming was occurring and stated that "*to embark on any vast experiment in the atmosphere would amount to gross irresponsibility.*"[8] In 1967, the US Department of Commerce released a report of the Panel on Electrically Powered Vehicles, "The Automobile and Air Pollution: A Program for Progress," which included a recommendation that emissions standards be set for vehicles,[9] which was done during the Ford administration through the nation's first Corporate Average Fuel Economy or CAFE standards.[10]

During the late 1960s and early 1970s, along with the enactment of the nation's seminal environmental statutes, Congress established the Council on Environmental Quality (CEQ), in part to develop and recommend policies that would promote environmental quality and meet the conservation, economic, health and environmental needs of the nation.[11]

Significantly, President Nixon was advised in September 1969 by Daniel Patrick Moynihan, then Nixon's counselor for urban affairs, that if climate change was not addressed, "It is now pretty clearly agreed that the CO_2 content will rise 25% by 2000. This could increase the average temperature near the earth's surface by 7 degrees Fahrenheit. This in turn could raise the level of the sea by 10 feet. Goodbye New York. Goodbye Washington, for that matter."[12]

In the same year, Eugene Peterson, chief of the Bureau of Land Management's Division of Basin Studies, published an article detailing results if CO_2 emissions were not reduced. He noted that the Arctic Ocean could have no ice for half of every year, affecting weather patterns in the Northern Hemisphere.[13]

Two months later, President Nixon wrote to his Director of OST regarding the UNESCO Conference on the Environment, "Man and his Environment: A View Toward Survival," on the obligation and responsibility of government to protect future generations from global climate change and polluted oceans.[14] Director DuBridge subsequently urged John Ehrlichman to pursue the development of non-internal-combustion engine vehicles so that they would be available when they were needed.[15]

The late 1960s and early 1970s were a period where the federal government began to see the energy policy link to climate change. The National Oceanic and Atmospheric Administration (NOAA) and Environmental Protection Agency (EPA) were established during this period, and CEQ issued its first annual report in 1970.[16] CEQ called for worldwide recognition of the long-term significance of human alterations of the climate system, affected by the burning and consumption of fossil fuels, which, if unchecked, would lead to "substantial melting of ice caps and flooding of coastal regions."[17] President Nixon set the tone for CEQ's first report in his 1970 State of the Union Address, saying:

> Restoring nature to its natural state is a cause beyond party and beyond factions. It has become a common cause of all the people of this country. It is a cause of particular concern to young Americans, because they more than we will reap the grim consequences of our failure to act on programs which are needed now if we are to prevent disaster later.

> Clean air, clean water, open spaces—these should once again be the *birthright* of every American. If we act now, they can be.[18]

President Nixon's administration similarly emphasized, in the United States' second Annual Plan for participation in the World Weather Program, that "In the longer term, *the quality of the atmosphere may well determine whether man survives or perishes.*"[19]

Dr. Edward E. David, Jr., who later became the president of Research and Engineering at Exxon, was director of the OST under President Nixon, and helped draft the administration's proposals on alternative energy and pollution control. On October 20, 1970, he wrote the following proposal to the White House: "The federal government must play a leadership role because these efforts are so large and so long-term that the fragmented power industry cannot be expected to do the job itself."[20] An OST memorandum made more detailed recommendations for research and development funding for new energy technology and stated: "[A]t present the funding and direction of most new pollution control technology and new methods of power generation must come from the Federal Government or not at all."[21] These statements recognize the central role of the federal government in shaping the nation's energy future.

In the latter months of 1970, Ehrlichman created the National Energy Subcommittee of the Domestic Council and the Council of Economic Advisors, which together with the OST "began to formulate a national energy policy."[22]

In my expert opinion, in the period shortly after President Carter took office in 1977, there was a growing sense of concern and indeed urgency within the federal government that fossil fuel burning was heating the planet and causing the climate to change in many ways that could be catastrophic, and that such climatic changes posed dangers to the lives and property of Americans, particularly its young people and future generations who would long live with the decisions made during the latter part of the twentieth century. This sense of concern deepened throughout the years of the Carter administration, as I will next describe.

2 THE JIMMY CARTER ADMINISTRATION (1977–1981)

Before taking up in detail the issues of the federal government's knowledge of both the climate danger and the opportunity to move away from fossil fuels, as well as its decision to continue pursuing fossil fuel energy, it is helpful to begin by recalling an important event in 1980 at which President Carter spoke. It illuminates these three conclusions well, in addition to my own personal engagement.

On Leap Day in 1980, President Carter's last full year in office, the president gave an important address at the Second Environmental Decade Celebration in the White House. Before the celebration, I had an opportunity to brief the president on the forthcoming "Global 2000 Report," which would be released later in July of that year. In his remarks at the celebration, President Carter noted:

> Just before lunch, Gus and I were discussing the long-term threats which just a few years ago were not even considered: the build-up of carbon dioxide; acid rain; the fact that 800 million human beings now suffer from lack of nourishment or disease; the fact that our population will increase 50 percent in the world by the end of this century. . . . These kinds of concerns affect you and me, and on some of them we've hardly begun to work on corrective action that might be proposed, much less accepted and implemented. This last decade, however, has demonstrated that we can buck the trends.[1]

Later in his address, President Carter listed eight "preeminent environmental challenges of the next decade" and included on that list "that we faced squarely such worldwide problems as the destruction of forests, acid rain, carbon dioxide buildup, and nuclear proliferation." President Carter also stressed the need for new energy directions. On his list of preeminent environmental challenges was "that we put this nation on a path to a sustainable energy future, one based increasingly on renewable resources and on energy conservation." He urged that "energy conservation has got to become a way of life" and that we develop solar and renewable energy sources, noting that "true energy security can only come from solar and renewable energy technologies."[2]

President Carter was proud to remind the environmental leaders in attendance that day of the 1978 National Energy Act and hoped, he said, that future generations would recognize it as leading a "massive and fundamental shift toward energy efficiency." He noted that his proposed 1981 budget called for spending over $2 billion on energy conservation, double the 1980 level.[3]

All that said and done, President Carter's address that day also noted, "It's important to pursue a broad range of alternative energy sources, including synthetic fuels," and he mentioned his "highly controversial" proposal for an Energy Mobilization Board to "eliminate unnecessary delays" in approving energy projects. Here, the president was referring in part to the energy development proposals in his famous "malaise" speech of July 15, 1979, where he proposed "the most massive peacetime commitment of funds and resources in our Nation's history to develop America's own alternative sources of fuel—from coal, from oil shale, from plant products for gasohol, from unconventional [natural] gas, from the Sun."[4] For instance, the federal synfuels program was created in 1980, had a rough life, and was terminated in 1985 as the oil market improved. The legislation to create the Mobilization Board never passed. Yet in some respects, Carter's proposals were merely ahead of their time. As I will describe subsequently, oil and gas markets are today awash with unconventional oil and gas thanks in large part to federal support and facilitation.

Much had already happened in the Carter administration before the president gave his Second Environmental Decade remarks in 1980. The White House, the Executive Office of the President, the Department of Energy (DOE), and several other agencies were certainly aware of the links and interactions among the three conclusions I list at the outset. Indeed, the preceding year, 1979, had been one of numerous reports flying about in the administration, as well as interagency dialogue and debate linking these three points. It was well understood by defendants DOE, the president, and the Executive Office of the President, for example, that to reduce carbon emissions to respond to the threat of climate change would require a new energy policy from the federal government. And, looking ahead, much regarding the climate issue was still to happen before the president's term ended. Still, it is notable that forty years ago the basic outlines of the federal government's response to the climate issue were already plainly visible: knowledge of the climate science, knowledge of alternatives to fossil fuels, and continued full-throttle support for fossil fuel development and use. This pattern would persist through subsequent administrations.

DURING THE CARTER ADMINISTRATION, THE FEDERAL GOVERNMENT'S AWARENESS OF CLIMATE SCIENCE AND THE LINK BETWEEN FOSSIL FUEL USE AND DANGEROUS GLOBAL WARMING AND CLIMATE CHANGE BECAME WELL ESTABLISHED

The President and the Executive Branch

In March 1977, in the early months of the Carter administration, a climate science workshop was sponsored by what would subsequently become the newly formed Department of Energy.[5] The important conclusions of that workshop speak to the knowledge of climate science in the federal government at that time. The Preface to the DOE report of the workshop summarized the conclusions of the participants:

> Implicit in all the panel reports is the acceptance of increasing atmospheric CO_2 content, well documented since 1958 and most probably the case since the industrial era began. That this rise has paralleled the increase in fossil fuel usage and is roughly equal to half the CO_2 liberated by industrial activity was also accepted. That fossil fuel usage is the sole cause of the increase, however, is under dispute: some significant fraction *may* be attributable to a decrease in the size of the biosphere. It also seems certain that there is enough fossil fuel

still available to raise manyfold the level of atmospheric CO_2, if the current models are anywhere near correct.

It was also accepted that carbon dioxide's radiative properties are well enough known to say that its increase will warm the lower atmosphere. But the interactions and feedback mechanisms within the climate system are so complex that considerable uncertainty exists about the magnitude of the effects.[6]

The president's science advisor during the Carter administration was Frank Press, the Director of the Office of Science and Technology Policy. Press wrote the president about the climate threat on July 7, 1977, in a memorandum copied to James Schlesinger, who would soon become the first Secretary of Energy. Press's memorandum summarized the threat:

> Fossil fuel combustion has increased at an exponential rate over the last 100 years. As a result, the atmospheric concentration of CO_2 is now 12 percent above the pre-industrial revolution level and may grow to 1.5 to 2.0 times that level within 60 years. Because of the "greenhouse effect" of atmospheric CO_2, the increased concentration will induce a global climatic warming of anywhere from 0.5° to 5°C. . . .
>
> A rapid climatic change may result in large scale crop failures at a time when an increased world population taxes agricultural limits to productivity. *The urgency of the problem derives from our inability to shift rapidly to non-fossil fuel sources once the climatic effects become evident not long after the year 2000; the situation could grow out of control before alternate energy sources and other remedial actions become effective. . . .*
>
> The present state of knowledge does not justify emergency action to limit the consumption of fossil fuels in the near term. However, *I believe that we must now take the potential CO_2 hazard into account in developing our long-term energy stragegy* [*sic*].[7]

Barely six months into the new administration, the president and his top energy advisor were apprised of the problem and its implications for the US energy system.

A report on "The Long Term Impact of Atmospheric Carbon Dioxide on Climate" prepared for the Department of Energy in April 1979, known as the JASON report, advised that the CO_2 influence on climate

APPROXIMATE DATES FOR DOUBLING THE CARBON DIOXIDE CONTENT OF THE ATMOSPHERE

ASSUMED RATE OF INCREASE IN USE OF CARBON FUELS	DATE OF DOUBLING OF CARBON DIOXIDE CONTENT OF ATMOSPHERE
Constant 4.3%	2035
Constant 3%	2050
4.3% 1978 - 2000	
3% 2000-	2040
2% 2000-	2050
1% 2000-	2060

Figure 2.1

Approximate dates for doubling the carbon dioxide content of the atmosphere. *Source*: MacDonald et al., *Long Term Impact*, 19.

was "widely accepted" and predicted levels of temperature increase through the middle of the twenty-first century.[8] The JASON report also predicted when atmospheric CO_2 levels would double for various rates of increase in fossil fuels (see figure 2.1).

Even as early as the Carter administration, there was not a debate over whether a clear climate threat existed, but there were uncertainties about how quickly different climate change harms would occur. One issue around which modeling results differed in the late 1970s was the amount of global average warming one could anticipate with a doubling of the CO_2 in the atmosphere. Carbon dioxide had been increasing, as shown from measurements at Mauna Loa, and was projected to increase (see figure 2.2, from the JASON report).

In May 1979, Frank Press asked the National Academy of Sciences to investigate this and related issues. The NAS convened a panel under the chair of MIT professor Jule Charney, and the panel met in July 1979. The concentration of CO_2 had risen that year to approximately 337 ppm.[9] The result was the famous Charney report. The Charney report was made widely available at the time both within and outside the administration and used government-sponsored and government-produced scientific research to support its findings. The well-known technical finding of the Charney report was as follows: "We believe, therefore, that the equilibrium surface global warming due to doubled CO_2 will be in the range 1.5°C to 4.5°C with the most probable value near 3°C." This warming, the Charney

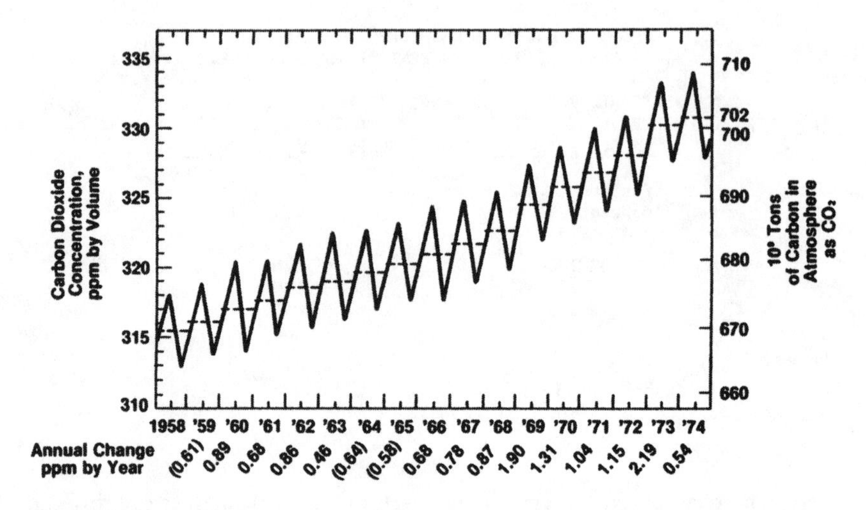

Figure 2.2

Atmospheric carbon dioxide concentration at Mauna Loa Observatory. *Source*: MacDonald et al., *Long Term Impact*, 6.

report concluded, "will be accompanied by significant changes in regional climatic patterns."[10]

The summary of the Charney report findings was particularly telling in its warning:

> The conclusions of this brief but intense investigation may be comforting to scientists but disturbing to policymakers. If carbon dioxide continues to increase, the study group finds no reason to doubt that climate changes will result and no reason to believe that these changes will be negligible. The conclusions of prior studies have been generally reaffirmed. However, the study group points out that the ocean, the great and ponderous flywheel of the global climate system, may be expected to slow the course of observable climatic change. *A wait-and-see policy may mean waiting until it is too late.*[11]

The DOE released a report in July 1980 on its "Summary of the Carbon Dioxide Effects Research and Assessment Program," which also reflected the scientific consensus:

> It is the sense of the scientific community that carbon dioxide from the unrestrained combustion of fossil fuels is potentially *the most important environmental issue facing mankind.* Current predictions call for a doubling of

> atmospheric carbon dioxide as early as the middle of the next century. Climate models, using these elevated levels, predict the possibility of significant dislocations in the global distribution of climate.[12]

This 1980 DOE report echoed the findings of the Charney report, noting that a doubling of atmospheric CO_2 was predicted to occur in the middle of the next century. Although uncertainties regarding timing and severity of warming persisted, in part due to the role of the oceans, the DOE report reflected an understanding of the dangerous impacts of the doubling of CO_2. (It is still the case that a doubling of atmospheric CO_2 is projected for the middle of this century.)

Congress

Congress was also aware of the climate and energy challenge during this early period. In a September 1977 report from the General Accounting Office (now Government Accountability Office) on future US coal development, the Comptroller General reported the following to Congress:

> [A] global warming of 1 degree to 2 degrees centigrade could cause serious repercussions on the earth's surface including shifting of wind circulation belts and redistributing temperature patterns and precipitation levels. Numerous secondary effects associated with these primary effects will also occur. . . .
>
> [T]he increased global temperature caused by rising concentrations of carbon dioxide may produce some melting of the polar ice caps, causing a sea level increase of tens of feet, gradually inundating coastal plains and low lands, and perturbation of marine biology. With continued growth in the use of fossil fuels, the effect of increased coal combustion on climatic conditions may become an important problem during the next 50 years.[13]

Proposed legislation on climate issues was introduced in Congress just as the Carter administration began. It became the National Climate Program Act of 1978 and its purpose was "to establish a national climate program that will assist the Nation and the world to understand and respond to natural and man-induced climate processes and their implications."[14] The act noted the great importance of climate factors and found that "an ability to anticipate natural and man-induced changes in climate would contribute to the soundness of policy decisions."

The agency I led in the Executive Office of the President, the Council on Environmental Quality, was mandated by the National Environmental Policy Act of 1969 to provide an annual report to Congress on environmental conditions, trends, policy actions, and results. We always saw the report as a vehicle for informing the public as well as the Congress, and we viewed it, among other things, as a warning mechanism for environmental threats. We issued four of these reports while I was a member and then chair of CEQ, starting in December 1977. Each was widely distributed and included contributions from other agencies, such as Office of Naval Research projections for atmospheric CO_2 increases that affirmed the work of climate scientists such as Charles Keeling.

In the CEQ's Eighth Annual Report to Congress in 1977, we wrote:

> If carbon dioxide levels increase, the amount of energy leaving the earth may decrease, resulting in higher temperatures in the lower atmosphere. The potential environmental consequences are many. . . .
>
> Deforestation, on the other hand, provides fuel, increases biological decay, and disturbs the soil—all of which increase CO_2 emissions. Deforestation also reduces the rate at which CO_2 is removed from the air. . . .
>
> If we use up the world's stores of fossil fuels at a rapid rate, the predicted CO_2 level will double by 2025 and reach a maximum of seven to eight times today's level by the year 2100. A doubling of the CO_2 level could cause a 2–3° C increase in average atmospheric temperatures. The most warming would occur at the poles. . . . A possible 2–3° C average temperature increase must be looked upon as a major global environmental threat—global temperatures over the past several thousand years have never fluctuated by more than about 1° C. . . .
>
> The global CO_2 problem is an issue which must be addressed in terms of its relevance to national energy policies. If further research confirms the hypothesis just described, then a programmed switch from fossil fuels to energy sources with no associated CO_2 emissions—such as solar power—may be imperative in order to limit global temperatures.[15]

The CEQ's Eighth Annual Report also included a graph (given in figure 2.3 here) showing the steady increase in CO_2 emissions, beginning in 1860.

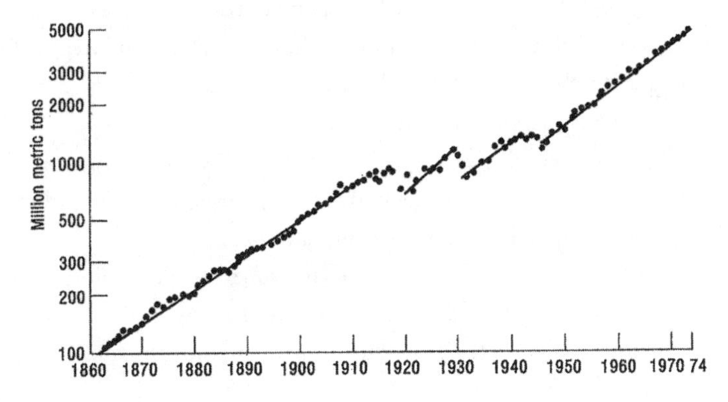

Figure 2.3
Estimated CO_2 global emission trends, 1860–1974. *Source*: Council on Environmental Quality, *Eighth Annual Report*, 190.

In the CEQ's Ninth Annual Report to Congress in 1978, we wrote to Congress:

> Global effects of carbon dioxide in the atmosphere—Combustion of fossil fuels, and especially coal, is increasing global atmospheric CO_2. This could induce climatic changes with potential for generating global sociopolitical disruption after 2025. It is urgent that we continue a strong research program to provide a sound basis for action no later than 1985. Because this problem is global in character, the United States should initiate a continuing international dialogue immediately.[16]

Building from the work we had done since the 1977 CEQ report, in the CEQ's Eleventh Annual Report to Congress in 1980, we strengthened our findings to Congress that the ongoing rising concentration of CO_2 from fossil fuel burning and deforestation would be disastrous for our natural systems and for humanity, particularly later generations of Americans. We wrote:

> Further, added to the major effects of fossil fuel burning, extensive loss of forests might aggravate the rising concentrations of CO_2 in the earth's atmosphere and thus eventually contribute to unprecedented human-caused changes in world climate. . . .

> There is a growing realization that the earth's atmosphere could be permanently and disastrously altered by human actions. The burning of fossil fuels and perhaps the cutting of forests without compensatory replanting are causing a steady, measurable buildup of carbon dioxide in the atmosphere that threatens widespread climate change. . . .
>
> A World Meteorological Organization (WMO) study group recently concluded that there is now little doubt that rising concentrations of carbon dioxide in the atmosphere will cause global warming. . . .
>
> A major contributor of CO_2 to the atmosphere is the combustion of fossil fuels. Most estimates of global energy use suggest that CO_2 concentrations could reach twice the pre-industrial level around the middle of the 21st century. . . .
>
> Possible climatic effects include changes in wind direction and speed, in ocean currents, and in precipitation patterns. If these large-scale climatic changes occurred, the socioeconomic impacts would be significant. If the warming continued long enough, polar ice could melt and sea levels would rise, forcing a gradual evacuation of heavily populated coastal areas. Agricultural patterns would change as well. In some regions existing agricultural infrastructure could become obsolete.[17]

In sum, based on my personal experience in the White House, these CEQ reports, and other historical evidence, by 1980, the US government was well aware both of the scientific findings that anthropogenic buildup of greenhouse gases in the atmosphere, principally carbon dioxide from the burning of fossil fuels, was occurring and that the projected and possible consequences of this buildup, if unabated, would be catastrophic.

A Key Energy and Policy Question Posed Early On Was What the CO_2 Target Should Be

In the 1977 National Research Council report "Energy and Climate: Studies in Geophysics," NAS explained, "The principal conclusion of this study is that the primary limiting factor on energy production from fossil fuels over the next few centuries may turn out to be the climatic effects of the release of carbon dioxide."[18] NAS explained that their "best understanding of the relation between an increase in carbon dioxide in the atmosphere and change in global temperature suggests a corresponding increase in average world temperature of more than 6°C, with polar temperature increases of as much as three times this figure."[19] Consequently, NAS advised that "in the light of a rapidly expanding knowledge and interest in natural climatic change, perhaps the question that should be addressed soon is, *'What should the atmospheric carbon dioxide content be over the next century or two to achieve an optimum global climate?' Sooner or later, we are likely to be confronted by that issue.*"[20] That key question posed in 1977 has been answered by scientists many times since (e.g., James Hansen[21]), and, indeed, we at CEQ offered an estimate as early as January 1981 (see later discussion), but the federal government has never adopted a ceiling for CO_2 buildup in the atmosphere.

DURING THE CARTER ADMINISTRATION, THE FEDERAL GOVERNMENT WAS WELL AWARE OF THE NEED TO SHIFT US ENERGY POLICY AWAY FROM FOSSIL FUELS TO RENEWABLES, EFFICIENCY GAINS, AND CONSERVATION

The Federal Government's National Energy Policy Included a Preliminary Push for Clean Energy

As early as the Frank Press memorandum to the president in July 1977 (a memorandum that reflected views widely held in the scientific community), there were calls for a new energy pathway that went straight to the White House.[22] As noted, although Press did not call for "emergency action," he advised President Carter that "we must now take the potential CO_2 hazard into account in developing our long-term energy strategy."[23]

The need to shift to nonfossil resources was not surprising news to President Carter or the federal government generally. Three years previously in 1974, Congress had passed the Solar Energy Research, Development, and Demonstration Act, which *inter alia* called for the creation of the Solar Energy Research Institute. In this legislation, Congress found that "dependence on nonrenewable energy resources cannot be continued indefinitely" and that "it is in the Nation's interest to expedite the long-term development of renewable and nonpolluting energy resources, such as solar energy."[24] Congress accordingly declared that it is "the policy of the Federal Government to (1) pursue a vigorous and viable program of research and resource assessment of solar energy as a major source of energy for our national needs."[25] Congress in 1974 was aware of the need for early commercialization of renewable technologies. The legislation notes that some solar technologies were "already near the stage of commercial application," and it called for the "demonstration of practicable means to employ solar energy on a commercial scale."[26]

President Carter moved rapidly in his first year in office to build on this legislation. In 1977 he created the Solar Energy Research Institute (SERI) and provided it with strong leadership and significant funding. (SERI was renamed in 1991 as the National Renewable Energy Laboratory.)

In April 1977, after just a few months in office, the Carter administration released the National Energy Plan, much of which would find its way into the National Energy Act of 1978. The official fact sheet on the president's program said: "The cornerstone of our policy is to reduce demand through conservation" and added the following statement: "Our energy problems have the same cause as our environmental problems—wasteful use of resources. Conservation helps us solve both at once."[27]

The National Energy Act of 1978 launched many of the energy efficiency initiatives needed to reduce fossil fuel use and foreshadowed others. It eliminated electricity rate structures that encouraged power use and, similarly, began the process of deregulating natural gas prices. It placed a tax on new gas-guzzling automobiles, created incentives for energy-saving investments, and imposed requirements for demand-side management by electric utilities, among other measures.[28]

At CEQ, we saw a need to share with the public what government knew about the near and long-term potential for solar and renewable energy. By April 1978 we had completed a report, "Solar Energy: Progress and Promise," which we made widely available within and outside of the federal government.[29]

The CEQ solar report defined "solar" to include renewables broadly and stated its goal on the first page of the Foreword:

> Despite the great potential of energy conservation, it alone will not be sufficient. We must also shift from oil and gas to other sources of supply. Yet,

the two most readily available, coal and nuclear power, are constrained by environmental and social problems.

It should not be surprising then that many of us in government and elsewhere are returning again to the questions: What can we reasonably expect of solar energy? And how soon?[30]

The CEQ solar report noted that "unlike coal, solar poses little risk to climate and creates little direct air pollution."[31]

Our conclusions at CEQ about the solar potential were more positive than we anticipated:

> Based on our review, the Council on Environmental Quality has reached some tentative conclusions about what would be reasonable goals for the United States in this vital area. No one's crystal ball works very well in examining energy futures, but based on available information and recognizing the uncertainties, we view the following goals as optimistic but achievable if we commit the necessary resources to them:
>
> - To make economically competitive over the remainder of the century a variety of solar technologies for the production of heat, electricity and biofuels.
> - To meet, by the turn of the century, a significant portion of our energy needs with solar energy. Although the actual contribution of solar energy will depend on an enormous number of decisions by the public and private sectors, we believe that under conditions of accelerated development and with a serious effort to conserve energy, solar technology could meet a quarter of our energy needs by the year 2000.[32]

The CEQ solar report in April was followed by President Carter's well-known Sun Day speech on May 3, 1978, in Golden, Colorado, the future home of Solar Energy Research Institute (SERI). President Carter began his remarks by noting that his energy proposals to Congress in 1977 had declared: "America's hope for energy to sustain economic growth beyond the year 2000 rests in large measure on the development of renewable and essentially inexhaustible sources of energy."[33]

He continued, citing our CEQ solar report in the process:

> We must begin the long, slow job of winning back our economic independence. Nobody can embargo sunlight. No cartel controls the Sun. Its energy

will not run out. It will not pollute the air; it will not poison our waters. It's free from stench and smog. . . .

The question is no longer whether solar energy works. We know it works. The only question is how to cut costs so that solar power can be used more widely and so that it will set a cap on rising oil prices. . . .

The Council on Environmental Quality recently estimated that we could meet as much as one-fourth of our energy demands for solar sources by the end of this century, and perhaps more than half by the year 2020. We must continue to make progress toward these goals.

The Department of Energy believes that photovoltaic cells can be competitive with conventional energy sources, perhaps as early as 1990. The Energy Department is working on many projects throughout this country, indeed throughout the world.[34]

President Carter announced major new funding for solar that day, but his major proposals for action came in his June 20, 1979, Solar Energy Message to the Congress, where he outlined, as the message says, "the major elements of a national solar strategy."[35] "The government-wide survey I commissioned concluded that many solar technologies are available and economical today. These are here and now technologies ready for use in our homes, schools, factories, and farms."[36]

After making the case to Congress for major coal use to replace oil, President Carter made an equally powerful case for solar and renewables. The goal President Carter announced of 20 percent renewables by the year 2000 was slightly below our recommended 25 percent target but reflected the CEQ's conclusions that a large shift to renewables was entirely possible with federal leadership.

In short, as early as 1978, the federal government was fully aware of and had begun acting on energy conservation and efficiency and alternative energy development policies, both in response to the oil crises that persisted through the 1970s, including the OPEC oil embargo of 1973–1974, and because alternatives to fossil fuels provided greater security and would have avoided the looming threat of catastrophic climate change. Notwithstanding Carter's and later recommendations regarding solar energy potential, US energy demand met with solar has consistently lagged far behind, as shown in figure 2.4.

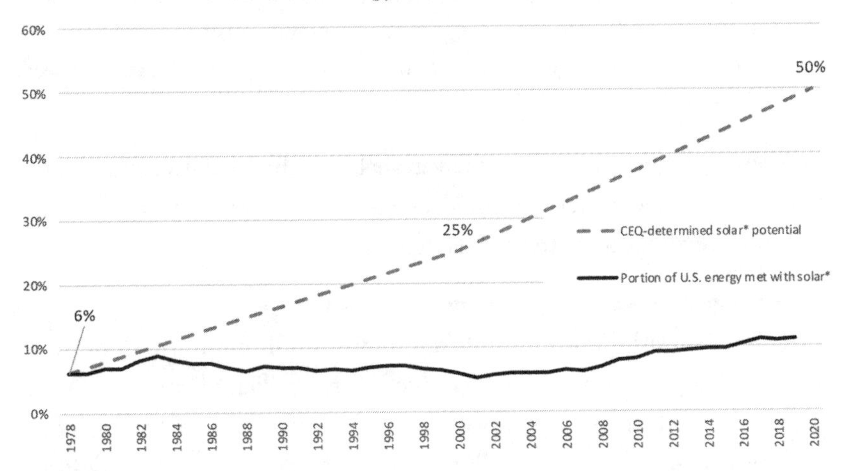

Figure 2.4

Portion of US energy that could be met with solar according to 1978 CEQ report: 25 percent by 2000; 50 percent by 2020. (Solar energy includes solar photovoltaic, hydroelectric, wind, and biomass–consistent with the definition in cited 1978 CEQ report.) Data from US Energy Information Administration, *Monthly Energy Review May 2018*, 2018. Ex. E-36; Council on Environmental Quality, *Solar Energy*, 6.

We at CEQ turned attention to the energy efficiency issue in 1979, releasing our report "The Good News About Energy." We undertook this energy efficiency report to explore the potential for achieving low energy growth in the United States and the implications of alternative energy growth paths for the economy and the environment. We ultimately found that energy consumption did not need to expand significantly in order to continue a healthy, expanding economy. We showed that by improving energy productivity and efficiency, we could keep energy consumption from increasing, by only 10–15 percent by the year 2000. We also found that "unforeseen developments [in technological improvements] seem far more likely to reduce energy growth than to increase it."[37]

In assessing the benefits of a low energy growth alternative, we noted that coal production could be about 50 percent less than business-as-usual and thus avoid the associated climate risks:

> Other longer-term environmental threats—equally difficult to quantify—will also be exacerbated by increasing energy use. The buildup of carbon dioxide,

> whose release is directly proportional to the amount of fossil fuels burned, is one. By absorbing a portion of the earth's outgoing radiation, carbon dioxide could lead to a long-term warming trend with potentially disastrous effects on the world's climate.[38]

Our report, which was made widely available to the Executive Branch, the Congress, and the media, presented a long list of potential energy efficiency measures beyond those already enacted.

Climate Enters the Policy Arena: 1979–1981

In 1979, the clash between the White House's call for climate protection and the federal government's fossil fuel energy policies became stark. As mentioned in the preceding, it was the year the often-cited Charney report was released and also the year President Carter gave his "malaise" address to the nation, with its emphasis on a massive new synfuels, coal to liquids, energy initiative. I was personally involved in bringing public and government attention to this entrenched conflict.

In May 1979, I met at CEQ with Gordon MacDonald, one of the United States' top atmospheric scientists, and Rafe Pomerance, then president of Friends of the Earth. They were seeking a stronger government response to the problem of global climate disruption. I promised to take the matter to the president and requested a reliable, scientifically credible memorandum on the problem from top scientists. The resulting July 1979 report, now more than four decades old, connected policy with scientific understanding of climate change, and was signed by four of our most distinguished American scientists—David Keeling, Roger Revelle, George Woodwell (lead author), and MacDonald.[39]

The contents of the 1979 report were alarming. The report predicted "a warming that will probably be conspicuous within the next twenty years," and it called for early action: "Enlightened policies in the management of fossil fuels and forests can delay or avoid these changes, but the time for implementing the policies is fast passing."[40]

Here are some further important excerpts from the 1979 report:

- Man is setting in motion a series of events that seem certain to cause a significant warming of world climates over the next decades unless mitigating steps are taken immediately. The cause is the accumulation of CO_2 and other heat-absorbing gases in the atmosphere.

- If we wait to prove that the climate is warming before we take steps to alleviate the CO_2 build-up, the effects will be well underway and still more difficult to control. . . . The potential disruptions are sufficiently great to warrant the incorporation of the CO_2 problem into all considerations of policy in the development of energy.
- Steps toward control are necessary now and should be a part of the national policy in management of sources of energy.
- The first element of any policy that offers the hope of being effective is conservation. Limitation of the rate of exploitation of fuels is possible. The rate is controlled currently by price, taxation, and regulation. It can be controlled as a matter of policy. All actions of government should be reviewed to determine effects on the total use of carbon-based fuels.
- It is our conviction that an appropriate reaction to the mounting worldwide squeeze on supplies of energy requires consideration of the CO_2 problem as an intrinsic part of any proposed policy on energy.[41]

The report was very clear on the urgency of bringing the climate issue into the formulation of national energy policy generally and the future of fossil fuels particularly. Unfortunately, later that same July the president would call for a major program to develop synthetic fuels (oil and gas) from coal and other hydrocarbons. The Woodwell–MacDonald–Revelle–Keeling report to CEQ contained a major warning about this policy. It strongly criticized the president's programs to increase coal production, stressing that synthetic fuels from coal and other hydrocarbons would release an estimated 2.3 times the amount of carbon dioxide per Btu compared to natural gas. The report made clear that the new synfuels policy the Department of Energy had developed for the president was inconsistent with protecting the climate system.[42]

This shot across the bow of the administration's plan to greatly expand domestic fossil fuel production was covered well in the press. The *New York Times* reported on July 10, 1979, that CEQ found the report to be historic and an important policy guiding document to bring energy policy in line with climate protection policy.[43] According to the *New York Times*:

> The new report has been sent to the President and other Administration leaders. Gus Speth of the Council of Environmental Quality, a White House advisory group, said, "The report is an extremely important perhaps historic statement." He added that he expected the report to be "very influential in government decision making." Mr. Speth also said that the report had shown that "the country needs to address the carbon dioxide issue squarely before going down the synfuels road." . . . Environmentalists have warned of potentially harmful effects from the rapid development of synthetic fuels, such as the release of toxics into the air and water, the rapid consumption of scarce water resources and devastation of the land for coal and shale mining. But little attention has been paid in the past by environmentalists and policy makers to carbon dioxide, which is odorless, colorless and poses no immediate threat to human health.[44]

Nonetheless, the 1979 report, the media coverage, and our efforts at CEQ did not deter the administration from announcing the synfuels program a few days later. Nonetheless, *the central policy issue—climate protection versus fossil fuel development—was joined in the policy arena for the first time*. DOE pushed back hard against CEQ and the climate scientists' 1979 report. As part of its push back, I recall that DOE produced a graph showing US fossil fuel use and CO_2 emissions growing so rapidly in future decades that the increment from synfuels development was simply dwarfed.

We were determined at CEQ to continue to press the matter of aligning energy policy with what the climate scientists were telling us was necessary to control climate change. We issued three subsequent reports to that end. Each received considerable media attention.

The first report, "Global 2000: The Report to the President," which CEQ produced over several years jointly with the State Department, was, as previously noted, released in 1980.[45] It was a "base case" analysis, looking twenty years ahead at future environmental and other conditions in 2000 if societies continued business-as-usual approaches. In its Principal Findings, the "Global 2000" report concluded that "atmospheric concentrations of carbon dioxide . . . are expected to increase at rates that could alter the world's climate . . . significantly by 2050."[46] Commenting generally on climate and other global-scale threats, the "Global 2000" report observed:

> Prompt and vigorous changes in public policy around the world are needed to avoid or minimize these problems before they become unmanageable. Long lead times are required for effective action. If decisions are delayed until the problems become worse, options for effective action will be severely reduced.[47]

The several volumes of the "Global 2000" report described the climate challenge and its links to fossil fuels and made a strong plea to prepare now for needed actions.

CEQ and the State Department also collaborated on a follow-up action plan to the "Global 2000" report, called "Global Future: Time to Act." Its recommendations for government action on energy and climate were forceful:

- An interagency task force should be established to chart a realistic path for achieving the goal of getting 20 percent of our energy from renewable energy by the year 2000. A national energy conservation plan, with near- and long-term sectoral goals, should be developed as part of the integrated strategy.
- The United States should ensure that full consideration of the CO_2 problem is given in the development of energy policy. Efforts should be begun immediately to develop and examine alternative global energy futures, with special emphasis on regional analyses and the implications for CO_2 buildup. The analyses should examine the environmental, economic, and social implications of alternative energy futures that involve varying reliance on fossil fuels, and they should examine alternative mechanisms and approaches, international and domestic, for controlling CO_2 buildup. Special attention should also be devoted to determining what would be a prudent upper bound on global CO_2 concentrations.[48]

In its coverage of "Global Future: Time to Act," on January 15, 1981, the *New York Times* noted that the report "followed the 'Global 2000 Report to the President,' which was issued last summer and warned that, without action to reverse them, current trends would lead 'to a more crowded, more polluted, less stable world' by the beginning of the next century."[49]

Our third, and most extensive, effort at CEQ to force a successful integration of energy and climate policy was not completed until around the time that President Carter unexpectedly lost the 1980 election. In "Global Energy Futures and the Carbon Dioxide Problem," we presented rigorously developed computer models of alternative energy futures and the climate risks associated with them. Based on this analysis, our recommendations to the federal government echoed the "Global 2000" report and were as follows:

- assign a high priority to incorporating the CO_2 issue into US energy policy planning;
- increase reliance on energy conservation and renewable sources of energy; and
- undertake new and expanded cooperative international efforts to address CO_2 issues.[50]

I summarized these conclusions in my Preface to the "Global Energy Futures and the Carbon Dioxide Problem" report:

> The CO_2 problem should be taken seriously in new ways: it should become a factor in making energy policy and not simply be the subject of scientific investigation. Every effort should be made to ensure that nations are not compelled to choose between the risks of energy shortages and the risks of CO_2. This goal requires making a priority commitment here and abroad to energy efficiency and to renewable energy resources; it also requires avoiding a commitment to fossil fuels that would preclude holding CO_2 to tolerable levels.[51]

Though the "Global Energy Futures and the Carbon Dioxide Problem" report was issued by a lame duck White House, it was widely distributed in Washington, DC, and elsewhere and garnered considerable media attention. I was quoted commenting on the report as follows in an article in the *New York Times* about the report:

> Gus Speth, chairman of the council, conceded that there was still some scientific uncertainty about the timing and effects of the carbon dioxide buildup

in the atmosphere. But Mr. Speth said that, given the magnitude of the risks and the fact that industrial countries were now formulating long-range energy plans, the carbon dioxide buildup must be considered in energy policy decisions. He said it would be too late to change course once the impact of the buildup began to be felt.[52]

The CEQ report recommended that a safe maximum level (or cap) for carbon dioxide in the atmosphere be established. In the late 1970s, it was thought that the CO_2 cap could be 50 percent higher than preindustrial levels, which would be approximately 420 ppm. (Today, climate scientists say that level is too high and that we have already exceeded safe bounds. See below.) The CEQ report addressed favorably a scenario that capped the buildup of CO_2 in the atmosphere at 50 percent above the pre-industrial level. On this matter, the *New York Times* article said:

> The level of carbon dioxide is currently estimated at 15 to 25 percent above pre-industrial levels existing around the year 1800. One recommendation of the report is that agreement be reached by industrialized nations on a safe maximum level for carbon dioxide in the air. It suggested a level 50 percent higher than that of pre-industrial times as an upper limit.

The CO_2 pre-industrial concentration is generally taken to be 280 ppm. A 50 percent increase would be 420 ppm. In July 2020, according to Mauna Loa data, the atmosphere's CO_2 concentration reached 414 ppm, and atmospheric CO_2 will likely reach 420 ppm, the 50 percent increase mark, in just a few years. Many, perhaps most, climate scientists now believe a 50 percent increase is too risky and would want the CO_2 buildup to stay below 25 percent. But it is noteworthy that the CEQ analysts were able to suggest an upper bound that was off only by a factor of two almost four decades ago.

President Carter was, I believe, prepared to tackle the climate issue in some meaningful way had he been reelected. But that was not to be. I think he would have asked his agencies for their ideas and plans on how to reduce US CO_2 emissions and would have led the United States on a very different path. It was not to be.

DESPITE THE EMERGENCE OF CLIMATE CONCERN AND THE PURSUIT OF ALTERNATIVE POLICIES TO FOSSIL FUELS, FOSSIL FUEL USE AND ITS EXPANSION WERE CONSISTENTLY PROMOTED IN MANY WAYS DURING THE CARTER YEARS

In parallel with its strong actions on energy conservation and its first-of-a-kind initiatives on renewables, and despite repeated warnings about the climate risks of fossil fuels, the Carter administration vigorously supported not only continued deep US reliance on fossil fuels but also a much larger reliance on coal in particular.

Coal use in the United States did indeed grow dramatically during and immediately after the Carter administration, while oil imports declined equally dramatically for a variety of reasons.

When Carter came into office in 1977, the country was still experiencing "gas lines" shock and stinging from its international oil vulnerability. In the decade before 1977, US oil imports had increased an astounding sixfold, and the OPEC oil embargo of 1973–1974 had deeply shaken American consumers and their politicians alike. The oil embargo gave rise to a major push to reduce oil imports by promoting "fuel switching"—shifting electricity generation from oil and natural gas to coal—and, some hoped, by using "synthetic" liquid fuels based on coal, tar sands, and oil shale. These heavier hydrocarbons produce more CO_2 per Btu when burned and are thus more harmful to climate. The key policy initiative in Carter's National Energy Program, a program put forth early in his first year in office, was, as the April 20, 1977, Fact Sheet said, "We must reduce our vulnerability to potentially devastating embargoes. We can protect ourselves from uncertain supplies by reducing our demand for oil, making the most of our abundant resources such as coal, and developing a strategic petroleum reserve."[53] The administration and the Congress pursued this objective with many policy initiatives, central to which was "fuel switching"—the shift in electrical power generation from oil and natural gas to coal. Fuel switching was among the policies strongly encouraged in the 1978 National Energy Act.

In addition to expanding coal leasing on federal lands, the Carter administration also joined with the Congress in promoting oil leases on the Outer Continental Shelf (OCS). In signing the Outer Continental Shelf Lands Act Amendments of 1978, Carter said, "This legislation will provide the needed framework for moving forward once again with a balanced and well-coordinated leasing program to assure that OCS energy resources contribute even more to our Nation's domestic energy supplies."[54]

At the end of his administration, Carter gathered information from throughout the federal government on what had been accomplished over the four years past. I recall working with CEQ staff to prepare our Memorandum to the President responding to this White House request. The result was Carter's January 16, 1981, State of the Union Annual Message to the Congress. It was an amazingly comprehensive 80-page document delivered to the Congress only in writing. In summarizing the administration's accomplishments, the first one listed by the president was: "Almost all of our comprehensive energy program have been enacted, and the Department of Energy has been established to administer the program."[55] Here, regarding legislative enactments, Carter was referring mainly to the National Energy Act of 1978 and the National Energy Security Act of 1980.

President Carter offered an overall summary of US energy policy that illustrates the power and control of the federal government over the national energy system, and the push for fossil fuel development notwithstanding the dire warning on climate change from ongoing fossil dependence.

> Since I took office, my highest legislative priorities have involved the reorientation and redirection of U.S. energy activities and for the first time, to establish a coordinated national energy policy. The struggle to achieve that policy has been long and difficult, but the accomplishments of the past four years make clear that our country is finally serious about the problems caused by our overdependence on foreign oil. Our progress should not be lost. We must rely on and encourage multiple forms of energy production—coal, crude oil, natural gas, solar, nuclear, synthetics—and energy conservation. The framework put in place over the last four years will enable us to do this.[56]

He then listed a number of specific accomplishments that underscore the commitment to fossil fuels both by the administration and by the Congress:

- Under my program of phased decontrol, domestic crude oil price controls will end September 30, 1981. As a result exploratory drilling activities have reached an all-time high;
- Prices for new natural gas are being decontrolled under the Natural Gas Policy Act—and natural gas production is now at an all-time high; the supply shortages of several years ago have been eliminated; . . .
- The Synthetic Fuels Corporation has been established to help private companies build the facilities to produce energy from synthetic fuels; . . .
- Coal production and consumption incentives have been increased, and coal production is now at its highest level in history; . . .
- In 1979 the Interior Department held six OCS [outer continental shelf] lease sales, the greatest number ever. . . .
- [T]he first general competitive federal coal lease sale in ten years will be held this month.[57]

The United States was about 90 percent dependent on fossil energy at the beginning of the Carter administration and close to 90 percent at the end; there was only a slight decline in fossil dependency of 2–3 percent. Meanwhile, the fossil fuel mix was shifting toward coal and away from oil imports. Total US energy use had declined slightly, and gross domestic product (GDP) had grown about 10 percent in real terms, so the overall energy efficiency of the economy had improved. A slow shift to renewables had started. It can be said that the energy policy of the Carter years was for the first time truly "all of the above"—and that included fossil fuels as the vast majority of our energy supply.

REFLECTIONS

It is clear looking back that, notwithstanding the unsupported climate denialism that has pervaded American politics off and on since the Carter years, and especially now, it is impressive how much was understood about climate change and the role of fossil fuels in causing it by the late

1970s, four decades ago. Enough was known, for example, to suggest a prudent upper bound for the buildup in the atmosphere of the principal greenhouse gas, CO_2, a limit that many reasonable people would fall on their knees to achieve today. For President Carter, as he said repeatedly, and for many of us inside the federal government, the path ahead was clear for our climate system and our nation's security and independence: energy efficiency, conservation, and renewables.

On Ronald Reagan's Inauguration Day, John Oakes, a member of the *New York Times*' editorial board, penned a warning for the incoming president in an article entitled "For Reagan, a Ticking Ecological 'Time Bomb.'" The *New York Times* piece accurately reflected the pivotal moment for our nation's energy and climate systems and the need to act swiftly to address the looming crisis. In it, Oakes wrote:[58]

> The rapid environmental degradation of this planet is a time bomb, as great a threat to both our national and our global survival as is the threat of nuclear annihilation. . . . The environmental crisis alluded to by Mr. Carter and described by a recent Government report, "Global 2000," is different in quality and degree from anything that has gone before in the history of the human race.[59]

Oakes went on to stress the climate threat: "The mad rush from oil to coal means more poisoned lakes from coal-produced acid rain, fouler air, and a prospective rise in world temperature (from accumulated carbon dioxide) that could dangerously raise the level of the seas." He pointed out that these threats "cannot be long ignored by the Reagan Administration." Oakes said this about solutions:

> What can we do about all this? In a sequel to "Global 2000," called "Global Future: A Time to Act," a Government task force headed by Gus Speth, chairman of the Council on Environmental Quality, has just proposed a string of recommendations to halt the slide into the environmental disaster that is sure to come if the new President and the new Congress fail to give it their urgent attention. This is a crucial issue—today. Like human life itself, once ecological systems are destroyed, they can never be recovered.[60]

Sadly, we know now that the Reagan administration and the Congress did indeed fail to give these issues their urgent attention.

EVERY SUBSEQUENT ADMINISTRATION KNOWINGLY CONTINUED DOWN THE FOSSIL FUEL ENERGY ROAD, FURTHER ENDANGERING THE CLIMATE SYSTEM

For the remainder of this expert report, I will describe how the federal government, under each presidential administration, (a) was made aware of the best and most current climate science; (b) had access to but largely ignored alternative policy pathways toward renewable energy development and energy efficiency; and (c) proceeded to pursue policies and take actions at the federal level that promoted the development and combustion of fossil fuels, thereby increasing carbon dioxide emissions and exacerbating the global warming crisis.

In 1976, the year President Carter was elected, the United States relied on fossil fuels for 91 percent of primary energy consumption. Over the ensuing decades, the seeds planted during the Carter administration regarding efficiency and renewables could have yielded a smooth transition toward an outstanding US climate performance and global leadership in climate action, which would have changed the trajectory of the atmospheric CO_2 levels. Instead, those decades saw only negligible action to actually reduce US emissions, with the result that US CO_2 emissions climbed by about 16 percent between 1975 and 2019. US fossil fuel consumption grew by even more. Fossil fuels still account for 80 percent of US primary energy consumption today.[61] During the period from 1980 to 2019, the carbon dioxide concentration in the atmosphere grew from 339 parts per million to 410 ppm, an increase of 21 percent. The United States was the largest contributor to this global change, and its failure to act undoubtedly influenced other nations.

This national energy policy of the last four decades is, in my view, the greatest dereliction of civic responsibility in the history of the Republic. And it is worse today than ever.

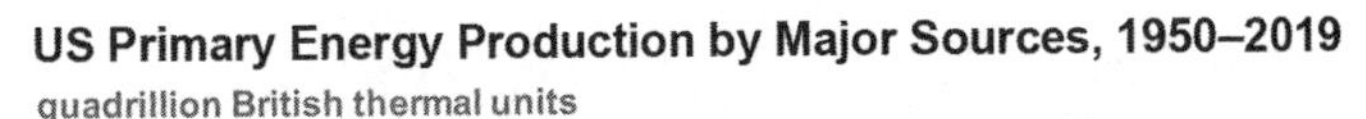

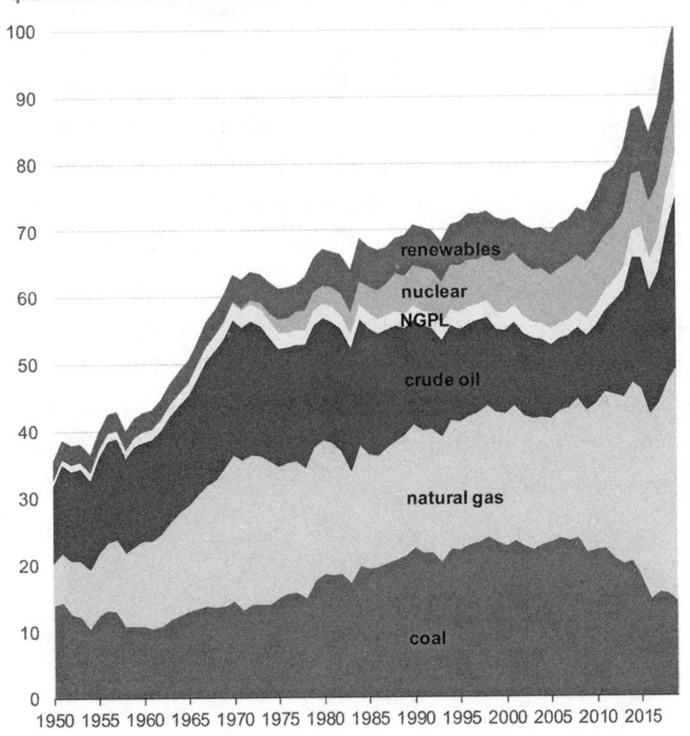

Figure 2.5

US primary energy production by major sources. Data from US Energy Information Administration, *Monthly Energy Review April 2020*, 2020, table 1.2. https://www.eia.gov/totalenergy/data/monthly/archive/00352004.pdf.

Though I did not serve in subsequent administrations as I had under President Carter, through my work at the World Resources Institute, at the United Nations Development Programme, and as Dean of the Yale School of Forestry and Environmental Studies, I have closely followed the federal government's positions and policies on global warming, climate change, and energy. What follows is my expert evaluation on what each administration knew about the science of global warming and each one's respective policy responses.

As the field of climate science and policy has evolved, the number of reports and testimonies provided to and by the federal government has grown exponentially. These reports have been produced, and testimonies have been delivered by scientists and others representing the federal government, research institutions, academia, and even private companies. There have been so many studies and reports over the past four decades that I couldn't possibly reference or discuss them all in this expert report. However, over the past four decades since leaving the Carter administration, I have kept apprised of and read hundreds of climate-related reports. I have also closely followed the escalating policy-related deliberations and actions on energy and climate. Accordingly, herein I reference some of the most important documents, testimonies, and events that illustrate how much the federal government truly did understand the risks of global warming. Similarly, my evaluation of policies and policy proposals will focus on what seem to me the most illustrative, representative, and impactful since President Reagan was inaugurated.

3 THE RONALD REAGAN ADMINISTRATION (1981–1989)

During the period of the Reagan administration, the field of climate science continued to mature and reveal ever more confidence about the reality of potential impacts. Despite the dire warnings of scientists, warnings that were presented to and even promoted by the federal government, the federal policies enacted during Reagan's eight years in office pivoted away from any real pursuit of renewable energy resources or efficiency programs and redoubled the US commitment to unrestrained fossil fuel development.

The commonly observed symbol of that pivot was when the Reagan administration removed from the White House roof the solar collectors put there by President Carter. *Scientific American* described the Reagan pivot in a nutshell:

> By 1986, the Reagan administration had gutted the research and development budgets for renewable energy at the then-fledgling U.S. Department of Energy (DoE) and eliminated tax breaks for the deployment of wind turbines and solar technologies—recommitting the nation to reliance on cheap but polluting fossil fuels, often from foreign suppliers. . . . And in 1986 the Reagan administration quietly dismantled the White House solar panel installation while resurfacing the roof.[1]

GOVERNMENT KNOWLEDGE

During the Reagan administration, the Executive Branch and Congress were both clearly aware of the link between fossil fuel combustion and the enhanced greenhouse effect, as well as the potential for catastrophic economic and human impacts of the resulting global warming and climate change. Between 1981 and 1988, research by governmental and nongovernmental scientists built upon and reaffirmed the early warnings issued by scientists in the 1970s. During these years, there was groundbreaking testimony before Congress by Dr. James Hansen and others, and their dire warnings about potential climate impacts were heard by the federal government and the American public. Meanwhile, President Reagan's own agencies produced a number of reports about global warming and developed strategic climate research plans, all while their budgets were slashed and resources drained.

Research and Reports on Climate Impacts

In June 1981, less than six months after President Reagan took office, the DOE convened a Workshop on First Detection of Carbon Dioxide Effects through its Carbon Dioxide Effects Research and Assessment Program.[2] The preeminent climatologists and meteorologists that attended and presented were described in the DOE's report on the workshop as "the best available scientists."[3] The purpose of the workshop was "to develop a research strategy that would provide the basis for early identification of the expected CO_2-induced response so that model projections of a warmer climate and of impacts on the biosphere can either be confirmed, rejected, or modified."[4] In its report on the workshop, the DOE noted the history of climate science awareness in the federal government, citing the 1965 report of the Environmental Pollution Panel of the President's Science Advisory Committee (PSAC), and the 1977 report "Energy and Climate" by the National Research Council that refined estimates of expected global temperature increases. The DOE was attempting to ensure that scientists and the government were prepared to identify the physical evidence of global warming as early as possible, "with the recognition that atmospheric CO_2 concentrations are increasing and that the consequent warming may have significant potential impacts on other climate elements and on important aspects of society, including the continued expansion of fossil fuel combustion." It is clear that President Reagan's DOE recognized the best available climate science projections at the time and considered the detection of impacts to be a priority supported by the federal government.

A year later, the DOE's Carbon Dioxide Effects Research and Assessment Program acknowledged the catastrophic risk of sea-level rise while examining the extensive research on the potential collapse of the West Antarctic Ice Sheet. The DOE's 1982 report stated:

> The serious consequences of the ~6 [meter] sea level rise that would occur in the event of a major shrinkage of the West Antarctic ice sheet would include flooding of all existing port facilities and other low-lying coastal structures, most of the world's beaches, extensive sections of the heavily farmed and densely populated river deltas of the world, and large areas of many of the major cities of the world, which are concentrated along coast lines.[5]

At the time, the DOE was the main federal agency researching CO_2, through the Assessment Program described in the preceding and DOE's Carbon Dioxide Research Division. For the first two years of the Reagan administration, DOE was producing and coordinating critical research on CO_2 emissions and potential climate impacts. However, in 1982, President Reagan cut the budgets of many of DOE's research offices, ignoring the recommendation of DOE's own Energy Research Advisory Board, which had recommended increasing funding for CO_2 research.[6]

In addition to the DOE, other agencies and councils linked to the federal government were conducting important research on CO_2 and climate change. For example, in 1982 the National Research Council issued a comprehensive report, affirming the government's longstanding knowledge of global warming, discussing the success of climate modeling, and explaining the role that the oceans play in absorbing most of the heat trapped by the excess CO_2 in the atmosphere.[7] The very first sentence of the report noted: "For *over a century*, concern has been expressed that increases in atmospheric carbon dioxide (CO_2) concentration could affect global climate by changing the heat balance of the atmosphere and Earth."[8] The report also stated:

> Despite the admitted existence of numerous uncertainties, the consensus on the nature and magnitude of the problem has remained remarkably constant

> throughout this long worldwide process of study and deliberation. Burning of fossil fuels releases to the atmosphere carbon that was extracted by ancient plants many millions of years ago. . . . Although questions have been raised about the magnitude of climatic effects, no one denies that changes in atmospheric CO_2 concentration have the potential to influence the heat balance of the Earth and atmosphere. Finally, although possibly beneficial effects on biological photosynthetic productivity have been recognized, no one denies that an altered climate would to some extent influence how humanity secures its continuing welfare.[9]

A National Research Council report published in 1983 stated: "While adverse consequences of 100 years from now are obviously less pressing than those of next year, if they are also of large magnitude and irreversible, we cannot in good conscience discount them."[10]

In 1983, President Reagan's EPA issued two seminal reports. In the first, "Can We Delay a Greenhouse Warming?," EPA projected an increase in temperatures of 2°C by 2040, a temperature increase that, in EPA's assessment, was guaranteed to produce substantial climatic consequences, including disastrous flooding.[11] Sea-level rise could increase by anywhere between 2 and 12 feet in the next century, and even the lowest possible rise will result in catastrophic damage to ports, transportation networks, and aquatic ecosystems and cause major changes in human migration. This warmer climate could affect everything from ecosystems to habitability of many areas of the world, and could impact relations between developed and developing nations. The report found that fossil fuels were responsible for most of the growth in the atmospheric concentration of CO_2 and thus stated eliminating coal combustion by 2000 and banning shale oil as the best policy options for delaying a 2°C temperature increase. EPA made clear that the risks are high with a "wait and see" policy.[12]

Importantly, the report did not endorse a 2°C warming target. On the contrary, it stated: "A 2°C temperature rise was selected because it represents a global warming significantly beyond the historical change for any 120 year period, and one guaranteed to produce substantial climatic consequences."[13] Also mentioned are the many previous instances of calls to action from Congressional hearings, newspaper editorials, and magazines, which the EPA believes are well founded, as the "concern about these changes far outweighs remaining uncertainties surrounding their exact nature and timing."[14]

The second EPA report, "Projecting Future Sea Level Rise: Methodology, Estimates to the Year 2100, and Research Needs," predicted 4.8 to 7 feet of global sea level rise by 2100, but said 11 feet could not be ruled out.[15] It found that CO_2 emissions were primarily caused by the combustion of oil, gas, and coal and that those emissions were consistently increasing.[16] EPA found that "future energy use and fuel selection will thus be the primary determinants of the rate of CO_2 emissions."[17]

In 1984, William Ruckelshaus, then EPA Administrator, made a powerful speech in Paris at the Organization for Economic Cooperation and Development. He acknowledged that the combustion of fossil fuels "has the potential for creating major climate changes," and that the failure to make a long-term commitment to address climate change and other environmental threats would lead to "a succession of unexpected and shattering crises."[18]

> There are some cherished values that resist being quantified or squeezed into monetary terms, but are no less real for that. Agents of democratic societies are responsible to the people, but we should remember that "the people" refers not only to the working majority that actually makes current decisions, and not even to the whole of the living population, but to those who came before us, who provided our traditions and our physical patrimony as nations, and to those who will come after us, and who will inherit what we leave behind. Decisions in the environmental arena often touch on this broader sense of public responsibility, and we cannot afford to lose it among the numbers.[19]

In another 1984 speech, Ruckelshaus referred to "the need to preserve our life support systems," and said

> I don't think our liberties are threatened in the next 90 days, but if we fail to improve our record in the realm of risk management, both within our societies and in the world as a whole, we will at least waste precious time and resources and at worst threaten all we hold dear.[20]

While the DOE's CO_2 research was hampered by President Reagan's budget cuts, its research did still continue. In 1985, scientists at the DOE's Lawrence Livermore National Laboratory edited and published a major, state-of-the-art four volume series for the Carbon Dioxide Research Division on the current state of carbon dioxide research. This series included volumes on "Projecting the Climatic Effects of Increasing Carbon Dioxide" and "Detecting the Climatic Effects of Increasing Carbon Dioxide," and was described by the editors as "an accounting" of what had been learned over the past eight years since the 1977 workshop described in the preceding. One of the volumes, "Atmospheric Carbon Dioxide and the Global Carbon Cycle," noted that CO_2 levels had varied in the last million years with a *high* point, during warm, interglacial phases, of 350 ppm.[21] An enormous amount had been learned; the "Projecting the Climatic Effects of Increasing Carbon Dioxide" alone was over 400 pages long, and is full of hard evidence of how much the federal government knew about the impacts of burning fossil fuels. For instance, in its preface, Michael R. Riches, Program Manager for the Carbon Dioxide Research Division at DOE, wrote:

> There is little doubt that the increasing concentration of atmospheric carbon dioxide (CO_2) has the potential to modify the Earth's climate. Increased global surface temperatures, altered precipitation patterns, and changes in other climatic variables could have substantial economic and social consequences. . . .
>
> Virtually all studies suggest that the increasing CO_2 concentration will significantly increase the global average temperature.[22]

Later in the volume, T. Webb III and T. M. L. Wigley, in their piece "What Past Climates Can Indicate About a Warmer World," wrote:

> The most significant effects of increased atmospheric CO_2 on climate will be manifest in regional changes of moisture and temperature patterns. . . . Changes in pressure patterns, both geographical and seasonal, will in turn affect rainfall, temperatures, and winds: all the meteorological variables that contribute to the overall climate at a given place.[23]

In the volume summary, Frederick M. Luther offered some dramatic conclusions:

As the climate warms, the amount of sea ice and the extent of snow cover will generally be reduced. . . . Feedbacks and interactions of many other types can also occur among elements of the climate system. . . . These model results suggest that global average temperatures will warm by 1.5 to 4.5°C once equilibrium is achieved after the CO_2 concentration has doubled. Continuing increases in the CO_2 concentration above those levels would warm the Earth still further. Such changes would be large in comparison to the decadal average temperature changes of the last 10,000 years, during which prolonged, global-scale variations have probably only rarely been more than about 1°C. . . . In summary, we have a sound qualitative understanding of the causes of the warming that is occurring and is projected to occur as a result of the increasing CO_2 and trace gas concentrations.[24]

In 1986, the National Oceanic and Atmospheric Administration prepared a report for the White House's Domestic Policy Council, the opening of which read: "We have become involved in a global climate experiment in which human activities . . . act as agents of inadvertent global change. . . . Recent and continuing increases in atmospheric carbon dioxide (CO_2) due to burning of fossil fuels can be thought of as a global climate experiment."[25] The report added, "We are reaching a technological threshold from which progress in understanding can be proportional to the investment of effort. This conclusion, combined with the need to formulate policy regarding current climate modification activities, deserves the attention of scientific and *governmental leaders who must determine and muster the needed resources.*"[26]

It was also in 1986 that eight senators sent a letter to Lee Thomas, administrator of the EPA, asking the EPA to conduct two studies on climate change.[27] The first study would examine the health and environmental impacts of climate change. The second study "should include an examination of the policy options that, if implemented, would *stabilize current levels of atmospheric greenhouse gas emissions.* This study should address: the need for and implications of significant changes in energy policy, including energy efficiency and development of alternatives to fossil fuel."[28]

In a separate letter, also in 1986, a group of senators asked the Office of Technology Assessment to come up with policy options that would stabilize and minimize greenhouse gases (GHGs) in the atmosphere, adding that they were "deeply troubled by the prospect of such a rapid and unprecedented change in the composition of the atmosphere and its implications for the human and natural worlds."[29]

It is noteworthy that the atmospheric concentration of CO_2 at the time these senators were asking the EPA and OTA to come up with concrete policy option to stabilize CO_2 at current levels was 347 ppm, just below the level where observations indicate that such significant impacts as loss of mass from ice sheets started accelerating. All three reports (two by EPA and one by OTA) were eventually prepared and submitted to Congress during the George H. W. Bush presidency (see later discussion of the reports in the next chapter).

DOE, EPA, NOAA, and other agencies had delivered an exhaustive, comprehensive accounting of the contemporary understanding of climate science during President Reagan's time in office. They had also explained that CO_2 wasn't the only greenhouse gas, but that other trace gas concentrations could contribute to the greenhouse effect and create a unified Greenhouse Gas Index.[30] DOE's Carbon Dioxide Research Division would add to this body of knowledge, officially publishing a Master Index to the state-of-the-art series in 1987 that again included dire warnings about the constantly updated state of climate science. The reports were conclusive that humans caused climate change and that temperature increases would be beyond anything seen in recent geological times.[31]

Groundbreaking Congressional Testimony by Dr. James Hansen and Others

Congress had heard from climate scientists before 1982, but earlier testimonies had largely covered what researchers knew about levels of carbon dioxide and predictions of future impacts. However, during a landmark hearing before the House Committee on Science and Technology on March 25, 1982, Dr. James Hansen and Dr. George Kukla first presented Congress with evidence of impacts. Hansen and Kukla had recently correlated increased CO_2 levels in the atmosphere with the shrinking of Antarctic ice and increases in worldwide sea levels. Thus, in spring of 1982,

Congress was alerted to precise and observable physical impacts of global warming resulting largely from the combustion of fossil fuels.

These observable impacts made modeling more accurate, and Dr. Hansen offered this projection related to sea-level rise:

> I would like to note that a smaller but still significant sea level rise is likely to occur in the coming decades even without collapse of the West Antarctic ice sheet. Just the thermal expansion of ocean water and the slow ice sheet melting, that we have evidence to be occurring, will probably raise sea level between 1 and 2 feet in the next 70 years, if the climate sensitivity is approximately of the magnitude estimated by the National Academy of Sciences committee chaired by Charney.
>
> A sea level rise of 1 to 2 feet is sufficient to cause large-scale beach erosion, intrusion of salt water into low-lying freshwater regions, and a large increase of damaging storm surges in coastal areas.[32]

In his written statement, Dr. Hansen warned, "Mr. Chairman, it is becoming increasingly clear that we should anticipate substantial climate change during the next several decades as a result of man's impact on the composition of the atmosphere."[33]

We know that the White House was aware of the testimonies at the hearings. A DOE report on the hearings and copies of testimonies were found in the Reagan Library files of Danny Boggs, who was a special assistant to the president at the time. The report on the hearing stated on the first page:

> The first witness, Professor Calvin, said that recent new evidence is giving us early warning signals that the greenhouse effect is, in fact, taking place, and we must not wait too long to act or it will be too late. The best way to avoid the problem is NOT put CO_2 in the air.[34]

Here is what Dr. Melvin Calvin told Congress in 1982:

> If we go on with a 50-terawatt fossil fuel strategy, the carbon dioxide emissions and the temperature changes are very large indeed. They go up to 4 or 5 degrees. A 4- or 5-degree rise in the global temperature means an enormous change in the agricultural pattern of the Earth, and if that happens within two generations of the human race, I do not think the human race can adjust to it

that fast; it is bad enough to adjust to it in several hundred years, to that big a change in agricultural patterns over the surface of the Earth. To do it in one or two generations, I think, is asking too much of mankind.[35]

In June 1986, Dr. Hansen testified again before Congress, along with other climate scientists. Senator John Chafee asked three scientists the question: "Do any of you believe that we need more scientific data before we could reach the conclusion that what is taking place now, if continued, will increase the temperature on the globe?" Dr. Hansen responded, "I don't think we need more evidence to say that." Dr. Rowland responded, "The fact that the greenhouse effect is working on the Earth, it seems to me, is perfectly straightforward." Dr. Watson responded, "No; I believe global warming is inevitable."[36] In his written statement, Dr. Hansen also stated, "Evidence confirming the essence of the greenhouse theory is already overwhelming from a scientific point of view."[37]

Two years later, in the summer of 1988, Congress would again hear from Dr. James Hansen during the Senate Committee on Energy and Natural Resource's hearing on the Greenhouse Effect and Global Climate Change. Dr. Hansen testified about the startling discoveries that had been made since his groundbreaking 1982 testimony.

I would like to draw three main conclusions. Number one, the earth is warmer in 1988 than at any time in the history of instrumental measurements. Number two, the global warming is now large enough that we can ascribe with a high degree of confidence a cause and effect relationship to the greenhouse effect. And number three, our computer climate simulations indicate that the greenhouse effect is already large enough to begin to effect [*sic*] the probability of extreme events such as summer heat waves.[38]

In addition to Dr. Hansen's testimony, the hearing included discussion of specific potential impacts, particularly droughts in western states and sea-level rise in coastal regions. It also included specific recommendations to Congress to promptly reduce fossil fuel use. According to Dr. Woodwell, then Director of Woods Hole Research Center:

What has to be done? There isn't any question. . . . We must reduce the use of fossil fuels on a global basis, a reduction of the order of 50 to 60 percent is probably appropriate, and the sooner the better. It is also true that cessation of deforestation on a global basis is completely appropriate to solve the climatic change problem and for many other reasons. . . . I'm not at all doubtful that such an objective is realistic. If we could establish that as a signal step in the process of reducing reliance on fossil fuel globally, I would think that we would have done one of the strongest and wisest things possible.[39]

As one might imagine, Hansen's 1988 testimony created waves in the media and elsewhere. The *New York Times* story was page one news: "Global Warming Has Begun, Expert Tells Senate."[40]

In September 1988, Donna Fitzpatrick, Undersecretary of DOE, testified before Congress that

> Evidence available from research to date from the Department's activities, from that of many other agencies and from other nations is sufficient cause for serious concern, even at the most optimistic end of the range of predicted results. This is of particular interest to the Department of Energy because U.S. fossil fuel use accounts for approximately 23 percent of the global total emissions of CO_2 resulting from combustion. . . .
>
> The prospects for future growth in the use of renewable technology appear especially promising as research continues to improve their efficiency, economics, and reliability. Renewable energy use can reduce carbon emissions and give developing countries attractive alternatives to the use of fossil fuels and further depletion of forests.[41]

At the same September 1988 hearing, Office of Technology Assessment Director John Gibbons discussed the urgency of developing policy responses to climate change. He stated:

> In global climate . . . the time constant is long—requiring decades to cause effects and almost indefinitely long for effects to reverse. And that's why it seems to me it may be foolhardy to procrastinate and apply Disney's law to the issue. That is, Disney's law says that wishing will make it so. I don't think we have a basis for wishing that this problem will fix itself. Therefore, what do we do? Well, one option is to cut back on our emissions. Another is to slow down the emissions and buy some time to understand things better. The third would be to offset the effects.[42]

The Reagan White House Council on Environmental Quality Warned of CO_2 Risks, Then Went Silent on Global Warming

In 1982, my successor at CEQ, Alan Hill, filed the CEQ's 12th Annual Report to the Congress for 1981, detailing the current state of the environment and spelling out President Reagan's recommendations for dealing with environmental issues. The CEQ report recognized the recent research, acknowledging that anthropogenic warming "may now be detectable" according to a NASA study, and that mean global temperature could increase "between 2.5 and 4.5 degrees Celsius by the end of the 21st century."[43]

However, the White House's recognition of climate science was short-lived, and the same report described President Reagan's plan to increase production of domestic fossil fuel resources. The next year, the CEQ's Annual Report for 1982 omitted any discussion of CO_2 and climate change, focusing almost exclusively on a blueprint for increased fossil fuel development. The 1983 CEQ Annual Report did not reference climate change, and the 1984 report only included a chart recognizing the increase in CO_2 emissions since 1958.[44]

In 1986, President Reagan received a letter from Senator Al Gore discussing the current climate change science and warning that "one of the most serious and long-term environmental problems facing the United States and the world is the greenhouse effect."[45] The assistant to the president responded to Senator Gore, "Your concern over the seriousness of this problem is being conveyed to the President" and "we will be pleased to review the option which you suggested to coordinate efforts to deal with this phenomenon."[46] Nevertheless, as discussed in the following section, the actions of President Reagan's administration indicate that inaction on climate change was not addressed.

GOVERNMENT ACTION

Enlarged Commitment to Fossil Fuels

Throughout President Reagan's time in office, while his own administration was producing and publishing groundbreaking research on the growing threat of greenhouse gas emissions from fossil fuels, the same administration firmly committed the country on a path of further extensive coal, oil, and gas development.

The president's first term was marked by rapid deregulation of fossil fuel energy markets and the opening up of lands for fossil fuel production, especially oil and gas. In 1981, President Reagan first introduced the elements that would form his national energy policy through the National Energy Policy Plan, which included dismantling environmental regulations and removing restrictions on production and leasing policies.[47]

The 12th Annual Report of the CEQ made clear the government's focus on expanding fossil fuel development, notwithstanding the threats posed by climate change. The report stated:

> Energy and minerals on public lands—both onshore and offshore—have been the focus of intense interest since the 1973–74 oil crisis. The Administration has placed these resources in central position in its energy policy. . . . "The Federal role in national energy production is to bring these resources into the energy marketplace, while simultaneously protecting the environment." . . . To carry out the President's program, the BLM Director announced a reorganization of the Bureau's functions . . . stating "One of our primary objectives is to increase the availability of federal lands and resources for energy and mineral development." . . . High priority is to be given to streamlining existing energy and mineral leasing programs; accelerating the development and implementation of new programs for leasing oil shale, tar sands, and Alaskan onshore oil and gas; increasing the availability of federal lands for exploration and development activities, particularly for oil and gas and strategic minerals.[48]

The report also states that the Interior Department increased oil and gas leases in 1981 by 36 percent, increased the acreage leased in 1981 by 152 percent, and ended the moratorium on oil and gas leasing on military lands.[49]

The Reagan administration opened up dramatically more federal lands for oil and gas development than in prior years, and sped up the sale of offshore oil and gas leases. In its 13th Annual Report, CEQ noted that "despite the presence in the offshore area of approximately two-thirds

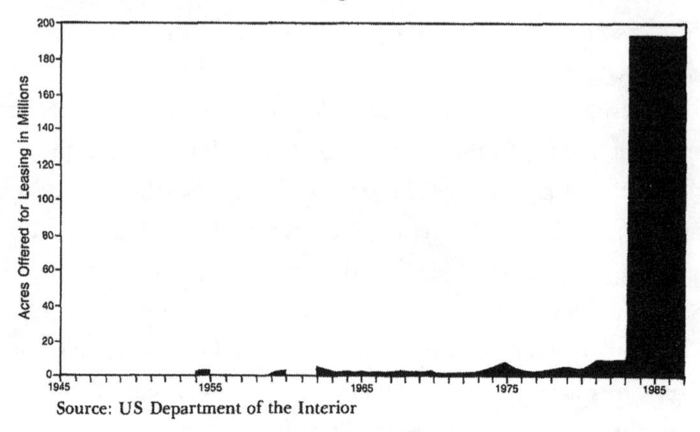

Figure 3.1
Offshore oil and gas leasing. *Source*: Council on Environmental Quality, *13th Annual Report*, 156.

of America's oil and gas resources, over the past 30 years little effort was expended to develop that potential."[50]

The figure that CEQ included in the report is telling (shown in figure 3.1).

Not only was the federal government opening up vast new acreage to potential oil and gas development, it spent hundreds of millions of dollars in research to facilitate use of these offshore areas for oil and gas exploration and development.[51] Between 1954 and 1980, the federal government had never offered more than 7.2 million acres of offshore areas for oil and gas development, or leased more than 1.9 million acres, in any one year.[52] However, by President Reagan's third year in office, the number of acres offered surpassed 119 million acres and the number of acres leased surpassed 6.5 million acres. By his fourth year, 154 million acres were offered, a twenty-one-fold increase from the previous peak in acres leased, and almost 7.4 million acres were leased, nearly a fourfold increase.[53]

The Reagan administration's work pertaining to onshore oil and gas development followed a similar trend. As CEQ noted: "In 1981, BLM [Bureau of Land Management] leased 150 percent more onshore acres than were leased in 1980. In 1982, leased acreage nearly doubled again, thus equaling in 1982 alone the number of acres leased for the entire period of 1977–1980."[54]

In his second term, President Reagan continued to maintain support for almost unrestrained fossil fuel development.[55] Reagan's antiregulation policies and price decontrol for oil and gas were executed by the heads of various federal agencies, including James Watt and Dan Hodel and the Department of the Interior (DOI).[56] It was also supported by the EPA, which, among other things, eased regulatory burdens for wastes from the exploration, development, and extraction of oil and natural gas and specifically exempted such wastes from the Resource Conservation and Recovery Act.[57]

A 1985 OSTP report, "Biennial Science and Technology Report to Congress: 1983–1984," emphasized science and technology that enhance fossil fuel recovery, for example, through hydraulic fracturing (fracking), but did not discuss the need for electric vehicles or high-speed rail or the need to increase the Corporate Average Fuel Economy (CAFE) standard. While noting some improvements in wind and solar energy, it focused on government research that would increase fossil fuels and largely left the research and development of renewable energy sources to the private sector.[58] Subsequently, the Reagan administration actually *increased* national auto emissions by undoing the fuel economy standards that had been put in place by prior administrations. President Reagan's Department of Transportation reduced the CAFE standard from 27.5 mpg to 26 mpg for 1986.[59]

CLIMATE CONCERN GOES GLOBAL

Meanwhile, as researchers in the federal government continued to sound the alarm about the threat of carbon emissions, momentum was building internationally for a cooperative response to climate change. As the longtime leader among the well-to-do countries, the United States was inevitably drawn into the climate issue internationally.

By the mid 1980s, the intellectual and policy leadership of the international scientific community, the environmental community in many countries, and the UN Environment Programme had paid off: A new and international environmental agenda had emerged—one that governments would be compelled to address collectively, or seem to address. The international pressure for major action on climate and other global-scale issues was becoming too strong to ignore.

The idea of an international climate convention entered international discourse at the important climate science conference held in Villach, Austria, in 1985. The Villach conference was the third of a series of meetings starting in 1980 that examined the impacts of CO_2. The conclusions and recommendations from the conference included:

- Climate change and sea level rises due to greenhouse gases are closely linked with other major environmental issues, such as acid deposition and threats to the Earth's ozone shield, mostly due to changes in the composition of the atmosphere by man's activities. Reduction of coal and oil use and energy conservation undertaken to reduce acid deposition will also reduce emissions of greenhouse gases. . . .
- While some warming of climate now appears inevitable due to past actions, the rate and degree of future warming could be profoundly affected by governmental policies on energy conservation, use of fossil fuels, and the emission of some greenhouse gases.[60]

The Villach results then fed quickly into the famous Brundtland Commission report—moving them from a scientific arena to a policy one. The 1986 Brundtland report, written by a United Nations-sponsored international commission, called for action to "minimize damage and cope with the climate changes, and rising sea level."[61] It urged international cooperation to achieve these ends "backed by a global convention if necessary." Over the ensuing few years, "if necessary" would become "as is essential." The United States participated as a member of the Brundtland Commission, and the organization I led at the time, the World Resources Institute (WRI), had a role in shaping the report.

The Villach conference and the Brundtland report led to the International Conference on the Changing Atmosphere convened in Toronto in 1988. (WRI also participated in the Toronto conference.) Affirming the need for targets and timetables for reducing emissions, the delegates at the conference suggested that global CO_2 emissions should be reduced by 20 percent below 1988 levels by 2005.[62]

In addition to starting the ball rolling on an international convention, the Villach and Toronto conferences had another major upshot, this one quite unintended. Many governments, including that of the United States, did not like the scientific community taking the reins on policy entrepreneurship. Thus, the Intergovernmental Panel on Climate Change (IPCC), an international body established in 1988 under the auspices of the United Nations Environment Programme and the World Meteorological Organization, was created (with US backing) for governments to assess anthropogenic climate change based on the latest science. Note the word *Intergovernmental.*

By the last year of Reagan's second term, advisors within his administration, including at the State Department, were advising Reagan to develop national climate policies and engage in international diplomatic efforts that were forming, such as the IPCC.[63] However, at the first IPCC meeting, the State Department made clear that while it was not refuting climate science, it was "premature" to negotiate any international agreement that "sets targets for greenhouse gases."[64]

Similar to the administration's approach internationally, on the national front the Reagan administration called for more research as opposed to taking action to confront the threats. In 1987, Congress passed the Global Climate Protection Act, stating that the "President, through the Environmental Protection Agency, shall be responsible for developing and proposing to Congress a coordinated national policy on global climate change."[65] However, as one 1987 memorandum from the White House stated, "OSTP has strongly opposed the Global Protection Act and consideration of policy actions on global climate change, arguing that the science is not understood well enough to formulate meaningful policies. Although the statute requires policy consideration of this issue, DPC [Domestic Policy

Council] action now could be premature before the science 'is ready.'"[66] In a January 1988 report to Congress on the greenhouse effect, President Reagan again emphasized the need for further research and interagency coordination, but offered no prescriptions for policy actions to reduce greenhouse gas emissions.[67] President Reagan also stated that his administration did "not plan to establish an International Year of the Greenhouse Effect as suggested in the language of Public Law 99–383."

Such was the political and social backdrop to the 1988 Presidential elections—a growing global call for international climate action, a Congress that was hearing repeated testimonies by Dr. Hansen and others on the forefront of climate research, and even agencies within the Reagan administration suggesting action to plan for and mitigate the potential impacts of warming. All the while, the federal government's policies were encouraging the enhanced production and combustion of fossil fuels.

4 THE GEORGE H. W. BUSH ADMINISTRATION (1989–1993)

During his presidential campaign, George H. W. Bush famously acknowledged the threat of global warming and promised action: "Those who think we are powerless to do anything about the greenhouse effect forget about the 'White House effect'; as President, I intend to do something about it."[1]

The first Bush administration's acknowledgment of the fundamental facts and implications of climate change is aptly summed up by comments made by Secretary of State James A. Baker III, a mere ten days after inauguration. Speaking to the Response Strategies Working Group of the Intergovernmental Panel on Climate Change (IPCC), Baker said:

> We can probably not afford to wait until all of the uncertainties have been resolved before we do act. . . . While scientists refine the state of our knowledge, we should focus immediately on prudent steps that are already justified on grounds other than climate change. . . . Solutions will be most effective if they transcend the great fault line of our times, the need to reconcile the transcendent requirements for both economic development and a safe environment.[2]

Not long after his swearing in, however, the actions of President Bush and his administration—both at home and, increasingly, abroad—began to undermine both these statements and the long-accumulating body of knowledge within the federal government and international community regarding the robustness of climate science and the risks of climate change. For the next four years, the first Bush administration would outwardly and quite vocally recognize the threat of climate change, the risks of ongoing high levels of greenhouse gas emissions, and the necessity of at least some action to counteract these risks. Yet the Bush administration continued to act in ways that would deepen US dependence on fossil fuels, foster high US greenhouse gas emissions, politically interfere with scientific research, appeal to alleged short-term "economic" justifications for inaction, downplay the significant short- and long-term consequences (economic or otherwise) of irreversible climate change, and slow international collective action to reduce greenhouse gas emissions. Defendants deeply understood the gravity of the emerging climate crisis and yet, when presented with pathways to decarbonization, continued to promote fossil fuel development and consumption.

GOVERNMENT KNOWLEDGE

The Administration Acknowledged Specific, Foreseeable Impacts from Climate Change and Sea-Level Rise

In his speech to the IPCC, quoted in the preceding, Secretary Baker channeled the thinking of Frederick M. Bernthal, who was then the Assistant Secretary of State for Oceans and International Environmental and Scientific Affairs and the Chair of the IPCC's Third Working Group. Assistant Secretary Bernthal submitted in a memo within the State Department that "while it is clear that we need to know more about climate change, prudence dictates that we also begin to weigh impacts and possible responses. We simply cannot wait—the costs of inaction will be too high . . . [T]he U.S. must take the lead in international efforts to address global climate change. Others look to us to do so because we are best equipped to understand the problem and develop solutions. We also contribute substantially to the problem."[3]

This wasn't the first warning of its kind to the federal government or the last of the warnings from Bernthal, who would be appointed by President Bush to lead the National Science Foundation (NSF) in 1990. Assistant Secretary Bernthal wrote to the Secretary of State, also in February 1989: "If climate change within the range of current predictions (1.5 to 4.5 degrees centigrade by the middle of next century) actually occurs, the consequences for every nation and every aspect of human activity will be profound."[4]

The EPA under the first Bush administration, with William Reilly as agency head, was in the process of recognizing and evaluating the great advances in climate modeling that had occurred in previous years, which introduced a new level of clarity as to the specific potential physical impacts of the greenhouse effect. Following the Congressional hearings on climate science in 1986, the Subcommittee on Pollution of the Senate Environment and Public Works Committee requested the EPA to conduct a study examining the potential health and environmental impacts of greenhouse gas-induced global warming. The EPA's resulting "The Potential Effects of Global Climate Change on the United States" report, published in 1989, provided an unprecedented and comprehensive evaluation of anticipated climate impacts, including on water resources, sea-level rise, agriculture, aquatic resources, air quality, and more.[5]

The language of the "Potential Effects" report regarding the broad climate situation was unequivocal:

> Recently, we have come to realize that human activities may, in the near future, produce effects powerful enough to overwhelm these natural mechanisms and dominate the changes of climate. By early in the next century, the planet's temperature may rise to a range never before experienced by our species, at a rate faster and to temperatures warmer than the Earth has experienced in the past million years. This anticipated temperature increase would be caused by an enhancement of the greenhouse effect.[6]

While the "Potential Effects" report language concerning the macro-level risks of anthropogenic climate change largely reflected that of previous government reports, the report is significant for the very high degree of confidence with which it described specific harms to specific regions from sea-level rise:

> A rise in sea level is one of the more probable impacts of climate change. Higher global temperatures will expand ocean water and melt some mountain glaciers, and may eventually cause polar ice sheets to discharge ice. Over the last century, global sea level has risen 10 to 15 cm (4 to 6 inches), and along

> the U.S. coastline, relative sea level rise (which includes land subsidence) has averaged about 30 cm (1 foot). Published estimates of sea level rise due to global warming generally range from 0.5 to 2.0 meters (1.5 to 7 feet) by 2100. . . .
>
> Although some wetlands can survive by migrating inland, a study on coastal wetlands estimated that for a 1-meter rise, 26 to 66% of wetlands would be lost, even if wetland migration were not blocked.[7]

Notably, the EPA also projected a time frame on the potential forest impacts, warning that "changes may begin in 30 to 80 years."[8] And, indeed, the impacts have begun in earnest today on the short end of the expected timeframe.

OTHER FEDERAL AGENCY REPORTS DEMONSTRATED INCREASINGLY SOPHISTICATED KNOWLEDGE OF CLIMATE RISKS

During the first Bush administration, the EPA was not the only agency to comment on climate risks. The Council on Environmental Quality (CEQ) discussed climate change in its annual reports. The Twentieth Annual Report, for instance, equivocated on the issue, stating that "emissions released to the atmosphere from fossil fuel combustion contribute to atmospheric degradation" but also that "the possible extent of climate change and its likely effects on the environment are both complex and controversial."[9] The CEQ's Twenty-Third Annual Report similarly sought to highlight uncertainties in the science, while noting that the United States' National Climate Change Action Plan was expected to reduce greenhouse emissions by 6–11 percent below business-as-usual levels by the year 2000. (US greenhouse gas emissions would in fact still increase in real terms over the period 1994–2000.)[10]

The General Accounting Office (GAO) also published a report, in March 1990, in which it explained all the ongoing research that DOE is conducting, including across 24 different programs, and stated in its "DOE Policy on Global Warming" section that "DOE has concluded that currently available information about the increase of carbon dioxide in the atmosphere is cause for serious concern even at the most optimistic end of the range of predicted impacts."[11] Later in the year, the General Accounting Office published a report, requested by Congress, that laid out what was currently known about climate change and greenhouse gas emissions and identified possible policy responses. "The average global temperature will increase by 3 to 9 degrees Fahrenheit over the next century, assuming a doubling of the effect of greenhouse gases," the GAO wrote.[12] "Even the lower of these estimates could be the most rapid temperature increase the earth has ever experienced."[13]

One year later, in winter 1991, the National Oceanic and Atmospheric Administration also published a public-facing report on "The Climate System" that spelled out the hard truths of climate science and the increasing probability of adverse impacts on society.[14] A chapter of the report titled "Water, Water Everywhere" stated:

> Rising ocean levels also threaten to flood low-lying areas and could create millions of refugees in Bangladesh and other countries. Urban centers like New Orleans, Bangkok and Venice may be unable to afford the costs of protecting themselves against the surge of high waves during storms.[15]

In (correctly) identifying that "no need could be more pressing, no mission of greater import to future generations" than "to anticipate future climate change and develop a rational program for protecting the environment," the report also showed an acute awareness of the particular temporal challenges inherent in responding to the climate crisis:

> The buildup of these and other gases has already strengthened Earth's greenhouse effect. But it may take several decades to feel the warming because atmospheric temperatures will rise significantly only after the oceans of the world have slowly warmed.
>
> The postponement may seem like an advantage, in that it gives us more time to prepare. However, the time lag could lead us to underemphasize the importance of the problem while we still have a chance to avert drastic climate change. In truth, we have already committed ourselves to some degree of warming, even if we could instantly halt the buildup of greenhouse gases in the atmosphere.

Whatever lies ahead, the world is accelerating its pace toward that unknown end. In the last three decades, the annual global release of carbon dioxide has doubled, reflecting a climb in the rate of fossil fuel burning and deforestation. As human population and economic activities continue to grow, carbon dioxide emissions could double again in the next three decades unless the nations of the world limit their consumption of fossil fuels.[16]

It is also worth noting that in 1990 the Global Change Research Act was passed into law requiring research on climate change and requiring the preparation of a report to Congress every four years on various climate change issues, including the environmental, economic, health, and safety impacts of climate change. The act stated that

> human activities, coupled with an expanding world population, are contributing to processes of global change that may significantly alter the Earth habitat within a few human generations.
>
> (2) Such human-induced changes, in conjunction with natural fluctuations, may lead to significant global warming and thus alter world climate patterns and increase global sea levels. Over the next century, these consequences could adversely affect world agricultural and marine production, coastal habitability, biological diversity, human health, and global economic and social well-being.[17]

The act brought together the research efforts of all the various federal agencies researching climate change under a single umbrella and coordinated effort that continues to this day.

THE FIRST BUSH ADMINISTRATION RECOGNIZED ALTERNATIVE POLICY PATHWAYS TO REDUCE CLIMATE RISK

In addition to the widespread and sophisticated knowledge of climate change held by multiple federal agencies during this first Bush administration, for at least a brief period federal agencies and Congress also seriously considered specific and comprehensive US policies for mitigating greenhouse gas emissions and reducing the risk of climate change.

Early recognition of the need to implement policies to mitigate the risks of climate change can be found in the 1989 memoranda from

Assistant Secretary of State for Oceans and International Environmental and Scientific Affairs Frederick M. Bernthal mentioned previously. In this February 9, 1989, memorandum, Assistant Secretary Bernthal stated:

> While it is clear we need to know more about climate change, prudence dictates that we also begin to weigh impacts and possible responses. We simply cannot wait—the costs of inaction will be too high. . . . Will focus on emissions reductions and encouraging reforestation for example.[18]

The memo went on to state that "the U.S. must take the lead in international efforts to address global climate change. Others look to us to do so because we are best equipped to understand the problem and develop solutions. We also contribute substantially to the problem."[19]

Assistant Secretary Bernthal reaffirmed this position—that policies to reduce greenhouse gas emissions could be implemented right away and justified in their own right, and the benefits of these would far outweigh the costs of inaction on climate change—in a February 27, 1989, memorandum to Secretary Baker:

> But a number of prudent measures could be taken that we would never regret, whether or not global warming ever occurs e.g., increased efficiency in energy use, global reforestation, and phasing out CFC production and use.[20]

Reflecting Assistant Secretary Bernthal's reasoning, the Department of Energy released a report in May 1990 titled "CO_2 Emissions from Coal-Fired and Solar Electric Power Plants."[21] The report found that "replac[ing] fossil fuels with renewable energy sources . . . is the only viable long-term strategy to provide power without curtailing growth" and that "appropriate technology" was then available to begin this transition.[22] The report went on to note:

> In view of the long lead time required for conservation and renewable technologies to be brought into the energy infrastructure on a large scale, policies that can be justified on their own merit and also reduce greenhouse-gas emissions should be initiated as soon as possible. There is a risk in delaying such actions since the costs of reducing CO_2 emissions are likely to increase if the urgency for their implementation should grow.[23]

These findings were reinforced in a September 1990 report entitled "The Economics of Long-Term Global Climate Change: A Preliminary Assessment, Report of an Interagency Task Force."[24] The Task Force included representatives from the Council of Economic Advisors; the Departments of State, Treasury, Commerce, Interior, Agriculture, and Energy; the Environmental Protection Agency; the Office of Management and Budget; and the Office of Science and Technology Policy. The Task Force's report made a number of findings in support of immediate, sector-specific policies to reduce greenhouse emissions and to transition the US energy system to renewables, including:

- Elimination of coal-mining jobs gradually over time does not necessarily imply increased general unemployment. . . . A shift to other energy sources would create jobs.
- Further increases in the CAFE standards are technically feasible and would likely reduce CO_2 emissions. . . .
- A number of changes in agricultural programs that would have other benefits can be expected to assist in reducing emissions of greenhouse gases. These include reducing commodity price support levels, encouraging additional tree planting, and conservation cross compliance.[25]

Additionally, a 1992 GAO report, "Energy Policy: Options to Reduce Environmental and Other Costs of Gasoline Consumption," discussed how a tailpipe tax, subsidies for alternative fuel, a higher CAFE standard, and other programs could be used to increase fuel efficiency and encourage the use of electric vehicles.[26]

While the first Bush administration's enthusiasm for concrete actions to address climate change would soon waver, federal agencies did develop comprehensive policy proposals for addressing greenhouse gases and stabilizing the climate in a way that would benefit the climate system and US economy. A major effort to assess policy alternatives was conducted by the EPA, as requested by the Subcommittee on Pollution of the Senate Environment and Public Works Committee. The EPA's "Policy Options for Stabilizing Global Climate" report, released in December 1990, "presents possible future scenarios of greenhouse gas emissions to the year 2100 depending on the level of response as well as many other independent factors," ultimately finding that the "results demonstrate that greenhouse gas emissions can be effectively reduced."[27] However, the report did note that "while it is not possible to stabilize greenhouse gas concentrations immediately" and "while humans may have already committed the earth to significant climate change during the next century, efforts undertaken now to limit the buildup of greenhouse gases in the atmosphere can dramatically reduce the rate and ultimate magnitude of such change."[28] Per the Congressional Committee's request,[29] EPA considered energy and emission pathways that would keep CO_2 levels at 350 ppm.[30]

The EPA modeled a number of different scenarios and evaluated the technological and economic feasibility of many emissions reductions efforts. The findings are summarized nicely:

> The adoption of policies to limit emissions on a global basis, such as simultaneous pursuit of energy efficiency, non-fossil energy sources, reforestation, the elimination of CFCs [chlorofluorocarbons] and other measures, could reduce the rate of warming during the 21st century by 60% or more. Even under these assumptions, the Earth could ultimately warm by 1–4°C or more relative to pre-industrial times. Extremely aggressive policies to reduce emissions would be necessary to ensure that total warming is less than 2°C.[31]

The EPA identified some of these policies and even evaluated their feasibility. For example, this report, 30 years ago, said, "50 mile per gallon automobiles are technically feasible with *currently available technology.* Further improvements could increase fuel efficiency to more than *80 miles per gallon.*"[32] "Policy choices and investment decisions made during the next decade that are designed to increase the efficiency of energy use and shift the fuel mix away from fossil fuels could slow the rate of buildup sufficiently to avoid the most catastrophic potential impacts of rapid climate change. Alternatively, decisions to rapidly expand the use of coal, extend the use of the most dangerous CFCs, and rapidly destroy the remaining tropical forests could 'push up the calendar,' accelerating the onset of a dangerous global warming."[33]

The EPA was not the only entity in the federal government to map out alternative scenarios during the Bush years. The Congressional Office of Technological Assessment (OTA) produced a report in 1991, as requested by Congress in 1988, that described policy options for reducing greenhouse gas emissions. The Director of the OTA wrote in the foreword that "this assessment focuses principally on ways to cut carbon dioxide emissions both in the United States and in other countries as well, although it does examine all greenhouse gases. . . . major reductions of carbon dioxide and other greenhouse gases will require significant new initiatives by the Federal Government, by the private sector, and by individual citizens."[34]

The OTA report identified the United States as the single largest contributor to carbon pollution and developed "an energy conservation, energy-supply, and forest-management package that can achieve a 20- to 35-percent emissions reduction. This package is labeled OTA's 'Tough' scenario. While difficult to achieve, major technological breakthroughs are not needed."[35] The report identifies "technical options" for lessening greenhouse gas emissions, and then spells out specific "policy instruments" available to government "to require or encourage a desired technical or behavioral response" (see figure 4.1).[36]

The 1991 OTA report made the following statements: "There is debate here as to whether and when a freeze or a 20-percent reduction in US greenhouse gas emissions could be achieved in the near-term. A 20-percent reduction in US CO_2 emissions would represent a 3-percent decline in current worldwide emissions of CO_2 and less than a 2-percent decline in current worldwide emissions of all greenhouse gases. More importantly, however, even if a 20-percent cut by all developed Nations could be achieved, it would not be enough to stabilize the atmosphere at today's level, let alone to reduce greenhouse gases to pre-industrial levels. To stabilize the atmosphere, the [IPCC and EPA] suggest, would require much more—up to an 80-percent global reduction in CO_2 emissions from current levels as well as significant reductions in the other greenhouse gases. To achieve this under the combined pressures of economic and population growth, nonfossil fuel technologies such as solar or nuclear power would be needed to replace much of today's fossil fuel use."[37]

Table 1-2—Policy Instruments To Reduce CO_2

	Taxes			Financial incentives			Marketable permits	Regulations		RD & D		Information	
	Energy tax	Carbon tax	Purchase tax	Tax incentive	Low cost loans	Direct payments		Performance standards	Building codes	R&D	Demonstration	Labels/ rating	Audits
Commercial buildings													
Thermal integrity	√	√	—	√	√	√	—	—	√	√	√	√	√
Appliance/lighting	√	√	√	√	—	√	—	√	—	√	—	√	√
Usage patterns	√	√	—	—	—	—	—	—	—	—	—	—	√
Residential buildings													
Thermal integrity	√	√	—	√	√	√	—	—	√	√	√	√	√
Appliance/lighting	√	√	√	√	√	√	—	√	—	—	—	√	√
Usage patterns	√	√	—	—	—	—	—	—	—	—	—	—	√
Transportation													
Small car/truck efficiency	√	√	√	√	√	√	—	√	—	√	√	√	—
Alternate fuels	—	√	—	√	√	√	—	—	—	√	√	—	—
Off highway[a]	√	√	—	—	—	—	—	√	√	—	—	—	—
Vehicle miles traveled	√	√	—	—	—	—	—	—	—	—	—	—	—
Manufacturing													
Efficiency	√	√	—	—	—	—	√	—	√	√	—	—	—
Recycling	√	√	—	√	—	√	—	—	—	—	—	—	√
Energy supply													
High to low carbon fuels	—	√	—	—	—	—	√	—	—	√	√	—	—
Renewable	√	√	—	√	—	√	√	—	√	√	—	—	—
Cogeneration	√	√	—	√	—	—	√	—	—	—	—	—	—
Efficiency, existing plants	√	√	—	—	—	—	√	—	—	—	—	—	√
Forests													
Recycling	—	—	—	√	—	√	—	√	—	—	—	√	—
Increased productivity	—	√	—	√	—	√	—	—	—	√	—	—	—
Afforestation	—	√	—	√	—	√	—	—	—	—	—	—	—
Biomass	√	√	—	√	—	—	—	—	√	√	—	—	
Food													
Farm inputs	—	√	—	—	—	√	—	√	—	—	√	√	√
Farm operation & efficiency	—	√	—	√	—	√		√	—	√	—	—	—
Food processing	—	√	—	—	—	—	=	—	√	—	√	—	√

[a]Heavy equipment, aircraft.

SOURCE: Office of Technology Assessment, 1991.

Figure 4.1

Policy instruments to reduce CO_2. *Source*: Office of Technology Assessment, *Changing by Degrees*, 14.

The OTA report goes on to conclude: "But it is clear that the decision to limit emissions cannot await the time when the full impacts are evident. The lag time between emission of the gases and their full impact is on the order of decades to centuries; so too is the time needed to reverse any effects. Today's emissions thus commit the planet to changes well into the 21st century."[38]

Unfortunately, the first Bush administration and Congress failed to act seriously on either the EPA or OTA plan, instead committing the nation to become even more dependent on fossil fuels.

GOVERNMENT ACTION

Despite the growing body of climate science and increasingly honed modeling that allowed for predicting specific impacts regionally, and despite the clear recognition of both the need for and feasibility of policies to reduce climate impacts, the federal government under George H. W. Bush continued to enact policies that would worsen greenhouse gas pollution, and this worked against the growing international scientific consensus and political momentum to take meaningful action, including international action, on climate change.

The Federal Government Kept Consideration of Climate Impacts Out of Federal Decision Making

On June 21, 1989, CEQ Director Alan Hill circulated a Draft Guidance to the heads of federal agencies, concluding that global warming was "reasonably foreseeable" and instructing agencies on how to prepare environmental impact statements and assessments that account for a projects' impact on climate change, as well as how climate change could later impact the federal project, suggesting that this is required under the National Environmental Policy Act (NEPA).[39] Hill had previously attempted to circulate a similar guidance at the end of the Reagan administration, but was prevented from doing so by intervention from the White House Counsel and the Domestic Policy Council.[40]

This time around, Hill again faced stiff opposition within the administration. In a June 23, 1989, memo, Acting Assistant Attorney General Donald A. Carr expressed concern about the effects of Hill's proposal on "litigation about the adequacy of environmental impact statements for continuing activities such as oil and gas development, timber sales, highways and coastal projects."[41] Carr "strongly urge[d] as a first step further deferral of the CEQ directive before things get out on the street, in the press and out of hand."[42] Hill was soon relieved of his position as head of CEQ, and David Bates, then Secretary to the Cabinet, drafted a one-page memorandum on July 14, 1989, that purported to "undo Hill's mischief" by advising agencies to abandon any efforts to address CEQ's draft guidance.[43] (Note that in 2010 the CEQ finally issued draft guidance on NEPA guidelines on how agencies must consider the impacts of their actions on global climate change. In 2014, after extensive public comment, revised draft guidance was issued, and in 2016, final guidance. However, this guidance was revoked by the Trump administration in March 2017.)

A 1990 GAO report subsequently confirmed that, notwithstanding campaign promises or the looming crisis, President Bush would not coordinate federal action on energy and climate policy to stop climate change:

> The President announced in February 1989 that he would issue an executive order on global climate change that would clearly define responsibilities of federal departments and agencies, as well as establishing effective coordination mechanisms. However, as of November 1989, the order had not been issued and its status was uncertain. Agency officials told us that they had not received clear guidance to direct the course of climate change activity.[44]

The GAO report further stated that "The administration has not tasked any agency with providing overall policy direction or leadership, nor has any agency acted as the administration's voice on global climate change" and that "NCPO's [National Climate Program Office] low placement in the department's executive echelon and a comparatively modest budget have hindered its effectiveness."[45] The first Bush administration never issued the referred to guidance or direction to federal agencies.

Administration Officials Undermined Agency Scientists and Domestic Confidence in Climate Science

As I have detailed in previous sections, the first Bush administration inherited decades of federal agency research, knowledge, and output in the area of climate science. Numerous agency reports and instances of congressional testimony by agency scientists had confirmed that warming of the atmosphere and resultant climate change were already occurring; that the cause was anthropogenic greenhouse gas emissions, foremost CO_2 emissions; that the adverse impacts from climate change would be numerous and potentially extreme; and that these impacts could be predicted with increasing accuracy and certainty. Given the implications of the evidence before it, it is saddening to observe that several prominent individuals within the first Bush administration rejected taking action and instead chose to undermine public confidence in the certainty of climate science, foster skepticism, promote contrarian viewpoints, and interfere with the work of government scientists.

One of the earliest examples of the first Bush administration running counter to the work of government climate scientists is the attempt by the White House Office of Management and Budget to alter testimony by then Director of the NASA Goddard Institute for Space Studies, Dr. James Hansen, before a Senate Committee chaired by then Senator Al Gore.[46] As Dr. Hansen stated in response to questioning during the 1989 Senate Committee hearing, the effect of the Office of Management and Budget alterations was to create confusion and internal contradictions in his testimony, and to cast doubt on the underlying science of climate change:

Senator Gore [to Hansen] In your statement you respond to our request for information on our scientific understanding of global climate models and our effort to determine which effects are pretty well understood and which effects are subject to change as we learn more about the models.

You respond by saying, among other things, that as the models improve and more evidence becomes available, it is not very likely that scientists will change their conclusion that increases in greenhouse gases will intensify drought in the middle and low latitude land areas, like the Midwest of the United States.

I am puzzled that you also say on that same point on page 4 of your statement that you want to stress that you do not really believe that and that as the computer models evolve, that conclusion will very likely evolve and should not be regarded as reliable.

I think I know the answer to the question I am about to ask you, but why do you directly contradict yourself in the testimony you are giving about this scientific question?

Dr. Hansen Let me first rephrase exactly what we said in that regard because when I discussed this with my scientific colleagues, the slight rephrasing makes a difference.

What I said was we believe it is very unlikely that this overall conclusion [*sic*] drought intensification at most middle and low latitude land areas if greenhouse gases increase rapidly, will be modified by improved models. Now, that is what I believe, and that is what I wrote.

The last paragraph in that section which seems to be in contradiction to that was not a paragraph which I wrote. It was added to my testimony in the process of review by OMB, and I did object to the addition of that paragraph because in essence it says that I believe that all the scientific conclusions that I just discussed are not reliable, and I certainly do not agree with that. . . .

Senator Gore Well, were there other parts of your testimony which they forced you to change?

Dr. Hansen The number of changes as these things go is actually not that large, but there was at least one other one which I think is worth mentioning. That was concerned with the—they added a sentence which says one point that remains scientifically unknown is the relative contribution of natural processes and human activities to the growth of trace gas climate forcings.

Now, I was able to get them to change the last part of that sentence to say "non-CFC climate forcings," because it is very clear that CFCs have no natural source. But, you know, even in the case of the growth of carbon dioxide and methane, it is pretty clear to scientists that in fact they are rising because of anthropogenic emissions.

I agree that the sentence is a scientifically correct sentence, but I would not have added that myself if I had not had it put in there for me. . . .

Senator Gore Dr. Hansen, were there scientists in OMB who ordered the change in your testimony?

Dr. Hansen I do not know them personally, so I really cannot say.

Senator Gore These are nameless, faceless individuals with whom you are dealing, is that correct?

Dr. Hansen Yes.

Senator Gore Sort of like members of the Science Politburo of the Bush administration. Well, I think this is an outrage of the first order of magnitude. I think that is evident.[47]

A magazine interview with then National Center for Atmospheric Research scientist Stephen Schneider[48] provided further indication that officials with the first Bush administration were interfering with government climate scientists and gatekeeping access to President Bush. When he was asked to comment on President Bush's position, he said:

> I suspect that his Chief of Staff John Sununu doesn't let a representative spectrum of climate scientists get near the Oval Office. The President apparently only hears unknowledgeable or extreme people. Neither [Bert] Bolin nor [James] Hansen nor myself have spoken with him. I doubt he gets balanced information on global warming or he'd join the [*sic*] Margaret Thatcher, and other "conservative" leaders, who counsel at least some immediate action to counter the greenhouse effect.[49]

Indeed, the first Bush administration's efforts to muddy the waters around the science of climate change were not limited to political interference with and censoring of government scientists. Another strategy, in concert with sympathetic members of the fossil fuel industry, was to foster support for climate science deniers, skeptics, and contrarians. For instance, in February 1991, President Bush received a letter from a group of scientists detailing a recent conference they had held, the purpose of which was to cast doubt on the consensus understanding of climate change and its risks.[50] By April 1991, Director of the Office of Science and Technology Policy D. Allan Bromley and White House Chief of Staff John Sununu had decided to invite several members to a meeting at the White House, with the apparent motive of seeking out and supporting scientists who "have been actively participating in the public debate on global warming from the point of view that perhaps the popular view—'The Popular Vision'—(IPCC science and/or potentially very negative effects) is wrong."[51] Bromley and Sununu appear to have convened the meeting in order to foster opposition to a soon-to-be released report, "Policy Implications of Greenhouse Warming," which Bromley and Sununu were concerned called for a more "activist" approach than was supported by the administration.[52]

White House Chief of Staff Sununu also appears to have been involved in soliciting support from the fossil fuel industry in undermining public consensus around climate science. On July 30, 1991, David Loer of the Minnkota Power Cooperative, Inc., wrote to Sununu as follows:

> Dear Mr. Sununu:
>
> On behalf of the 80,000 customers served by our rural electric power system, I want to thank you for the strong position you have taken on the global warming issue.
>
> Your skepticism regarding carbon dioxide as a cause of global warming is a "breath of fresh air" to those of us who are very concerned about the consequences of adverse action or legislation dealing with this issue.
>
> In our research, we have also found a number of credible climatologists who are not agreeing with the catastrophic global warming theory. We need to find ways for them to be more involved in dealing with this important issue.
>
> We know there is an extreme amount of pressure on you and other staff in the Administration to convince you that there needs to be limits placed on carbon dioxide emissions. We appreciate your recognizing the severe economic impact of such action.
>
> Minnkota Power Cooperative, Inc., owns and operates only North Dakota coal-fired electric generating plants. Not surprisingly, we are hoping that your position on this issue prevails.

> Is there anything that we can do to assist you or your position on the global warming issue? Please call on us—we would like to help.[53]

Sununu's response was brief, but telling of his support for the industry position:

> Dear David,
>
> Thank you very much for your note. My only suggestion to you on how you might assist us on the global warming issue is to encourage your colleagues in the industry to provide assistance and support to a credible climatologist who understands the complexities of the issue.
>
> We really *do* need help in getting this important message out.[54]

The Federal Government Undermined and Watered Down International Climate Science Assessments

During the first Bush administration, the United States exercised heavy influence over the shape and content of international scientific assessments. It largely did this behind-the-scenes while continuing to take a cooperative and comparatively positive stance in international fora. Thus, President Bush in his 1990 address to the IPCC stated:

> We all know that human activities are changing the atmosphere in unexpected and unprecedented ways. Much remains to be done. Many questions remain to be answered. Together, we have a responsibility to ourselves and the generations to come, to fulfill our stewardship obligations.[55]

As the main liaison from the first Bush administration to the IPCC, Assistant Secretary Frederick M. Bernthal helped craft the policy section of the panel's first report, "Climate Change: The IPCC Scientific Assessment," which was published in August 1990.[56] A particularly conservative block of important White House advisors—including chief of staff John Sununu and science advisor Allan Bromley—insisted on highlighting uncertainties in climate science, introducing the purported benefits of global warming while downplaying the central dangers of fossil fuel emissions and CO_2.[57] This same group of advisors also took the extraordinary step of directly writing to Bert Bolin, chair of the IPCC, and repeatedly demanding that changes be made to drafts of the IPCC First Assessment Report's "executive summary" and "overview and conclusions" sections and threatening to withhold US support for the report if the changes were not made.[58]

Bush's White House advisors, often adversaries of EPA's William Reilly, knew that if they accepted the international consensus on climate science, they would be forced to accept needed remedial action. As Bromley stated in a memo to Sununu on June 14, 1990:

> I am also concerned that if we appear to accept IPCC [Working Group I: Scientific Assessment of Climate Change] as ultimate truth then to some extent we have been effectively coopted into accepting the recommendations of the IPCC [Working Group II: Impacts Assessment of Climate Change] and [Working Group III: The IPCC Response Strategies] which may well be substantially different from what we had in mind.[59]

Indeed, as explained in the following sections, the United States would use its successes in watering down the conclusions of the IPCC First Assessment Report as a negotiating tactic in weakening the commitments and language of the 1992 framework convention.

The Federal Government Weakened the Language of and Resisted the Inclusion of Targets and Timetables in the United Nations Framework Convention on Climate Change

During his presidential campaign, then Presidential candidate George H. W. Bush announced his plans for his administration to spearhead international action on climate change and other environmental issues of global concern. A State Department memorandum quoted candidate Bush stating:

> In my first year in office, I will convene a global conference on the environment at the White House. All nations will be welcome—and indeed, all nations will be needed. . . . The agenda will be clear. We will talk about global warming. We will talk about acid rain. We will talk about saving our oceans and preventing the loss of tropical forests. And we will act.[60]

International developments over the previous years—including the 1985 Villach and 1988 Toronto conferences (in which the World Resources Institute [WRI] participated) and the Brundtland report—laid the groundwork for a climate convention, and by late 1990, the United Nations General Assembly had approved the start of negotiations leading to such a convention.

Yet in spite of Bush's early call for action, his administration set about using the United States' influence to weaken international action on climate change. Even before formal negotiations began, the first Bush administration worked to limit the strength of any agreement. While other nations, including most of Europe, argued for binding targets to protect the climate system, the United States would not commit to firm targets or timetables.[61] This opposition, based on "a volatile mixture of ideology and politics" rather than a "rational assessment of the national interest," was instrumental in ensuring that no binding targets were agreed to in the 1992 climate convention, a major setback.[62]

A seminal moment featuring this Bush opposition occurred at a high-level meeting of world environmental ministers in Noordwijk, The Netherlands, in November 1989. Many governments and environmental leaders were hoping the meeting would recommend binding targets and timetables for reduction of greenhouse gas emissions, reductions that would become part of the climate convention then being developed. It was not to be. Nathaniel Rich describes the results in his book *Losing Earth: A Recent History*:

> When, close to dawn, the beaten delegates finally emerged [from negotiations], [environmental advocates] Becker and Pomerance learned what had happened. Bromley [the president's science advisor], at the bidding of John Sununu [the president's chief of staff] and with the acquiescence of Britain, Japan, and the Soviet Union, had forced the conference to abandon the commitment to freeze emissions. The final statement noted only that "many" nations supported stabilizing emissions—but it did not indicate which nations, or at what level, or by what deadline. And with that, a decade of excruciating, painful, exhilarating progress turned to air. . . . In Washington, [Senator] Al Gore mocked Bush on the floor of the Senate: for all the brave talk about "the White House effect," the president was practicing the "whitewash effect." The United States had proved itself "not a leader, but a delinquent partner," said [Senator] Timothy Wirth. "I am embarrassed."[63]

The United States was largely isolated in several other negotiating areas, including whether to incorporate the "precautionary principle" in the language of the convention, the level of financial assistance provided to developing countries to offset the costs of reducing greenhouse gas emissions, and the degree of technology transfer between parties on noncommercial terms.[64]

Despite successful US efforts to weaken the content of the UN Framework Convention on Climate Change (UNFCCC), the convention was adopted on May 9, 1992, signed by the Bush administration in June 1992 at the Earth Summit in Rio de Janeiro, and later ratified by the US Senate in October 1992. I was there at Rio, saw it happen, and recall a certain pride with, at last, a feeling of hope for the future: At least something had finally happened. Despite being greatly weakened by the Bush White House, the treaty had some positive features and was a start at international cooperation. It provided the framework, we hoped, for the adoption of a strong implementing protocol, one with binding targets and timetables. (On a personal note, I would add that over the last half century we advocates for climate action have been forever hopeful, despite setback after setback.)

In ratifying the climate convention, the federal government was not on autopilot. Two aspects of the US adoption of the climate convention in particular underscore the seriousness of the United States' undertaking. Simultaneous with signing the climate convention, the United States rejected a parallel, but less far-reaching, convention aimed at protecting global biological diversity. And, second, the Senate's ratification of the agreement has proven a rarity indeed. The US Senate is a veritable graveyard of unratified multilateral environmental agreements going back at least to the 1982 Law of the Sea. The potential import of the 1992 climate convention was therefore unmistakable, as was its meaning for US action. In the end, despite the serious omissions, much of the language of the convention was quite promising.

The preamble to the UNFCCC states that the parties to the convention are "*Determined* to protect the climate system for present and future generations."[65] The parties, comprising now essentially every nation on earth (including the United States), expressed their concerns:

that human activities have been substantially increasing the atmospheric concentrations of greenhouse gases, that these increases enhance the natural greenhouse effect, and that this will result on average in an additional warming of the Earth's surface and atmosphere and may adversely affect natural ecosystems and humankind.[66]

The UNFCCC recognized "the need for developed countries to take immediate action in a flexible manner on the basis of clear priorities, as a first step towards comprehensive response strategies at the global, national and, where agreed, regional levels."[67] It also noted remaining scientific uncertainties, stating that "there are many uncertainties in predictions of climate change, particularly with regard to the timing, magnitude and regional patterns thereof."[68]

The most cited language in the convention comes in stating the convention's overall objective, "to achieve, in accordance with the relevant provisions of the Convention, *stabilization of greenhouse gas concentrations in the atmosphere at a level that would prevent dangerous anthropogenic interference with the climate system.*"[69]

All parties further committed to "Formulate, implement, publish and regularly update national and, where appropriate, regional programmes containing measures to mitigate climate change by addressing anthropogenic emissions."[70] Specifically, the developed country signatories committed to "adopt national policies and take corresponding measures on the mitigation of climate change, by limiting its anthropogenic emissions of greenhouse gases and protecting and enhancing its greenhouse gas sinks and reservoirs."[71]

Throughout, the treaty stresses the need to base decisions on the best science. For example, parties agree to

> Promote and cooperate in scientific, technological, technical, socio-economic and other research, systematic observation and development of data archives related to the climate system and intended to further the understanding and to reduce or eliminate the remaining uncertainties [and] . . . Promote and cooperate in the full, open and prompt exchange of relevant scientific, technological, technical, socio-economic and legal information.[72]

This history and these treaty commitments underscore the full awareness by the federal government of (1) the climate threat to current and future generations and (2) the need to take concrete actions to "prevent dangerous anthropogenic interference," especially action by the developed countries such as the United States. Actions by the Bush administration in signing the UNFCCC and by Congress in ratifying it, as well as the language of the UNFCCC itself, should have signaled the beginning of ongoing federal leadership at home and abroad to reduce GHG emissions, for the quoted treaty language endorsed by the Executive Branch and Congress was strong indeed in many respects. But the commitments the federal government made in 1992 in fact were not honored.

The Federal Government Introduces a New National Energy Strategy That Encourages More Fossil Fuel Development

In contrast to the optimism embodied by its signing of the UNFCCC, the energy policy legacy of the first Bush administration was instead a cause of further endangerment of the climate system.

During his first year in office, at the signing of the Natural Gas Wellhead Decontrol Act of 1989, President Bush announced that the DOE would develop a new National Energy Strategy.[73] The strategy, which was ultimately released in February 1991, openly acknowledged the risks of global warming, even claiming that the plan would reduce greenhouse gas emissions. However, as implemented, emissions continued to rise, and the National Energy Strategy encouraged such a build-out of infrastructure for oil and gas that it ensured a longer future of greenhouse gas pollution.[74]

Although the National Energy Strategy included a broad focus on using energy more efficiently, it supported increases in domestic production of oil, gas and coal,[75] and made "little effort to reflect in energy prices all the costs to society of obtaining and using energy, such as the adverse environmental consequences of relying on fossil fuels."[76] Specifically, it called for opening the Arctic National Wildlife Refuge (ANWR) and other areas of the Outer Continental Shelf (OCS) to oil production, and to "implement oil and gas tax incentives," to "deregulate oil pipelines," and to "increase production of California Heavy Oil."[77] It also set out to "provide regulatory incentives to offset financial risks in commercial deployment of new clean coal technology" and to "reduce the cost, investment risks, and environmental impacts of producing liquid fuels from coal."[78] DOE had already sought large increases in funding for so-called "clean coal" research and development as early as 1989.[79] The DOE was prioritizing research and development that would allow for more fossil fuel production, rather than the development of needed renewable energy sources. For example, for the 1990 fiscal year, DOE requested $575 million, a 242 percent increase over the fiscal year 1989, for clean coal technology. Meanwhile, DOE's funding request for conservation and renewable programs were reduced for fiscal year 1990.[80]

Just four months after signing the framework convention on climate change and promising the world at the Rio Earth Summit "forceful action" on climate change and international cooperation,[81] on October 24, 1992, President Bush signed the Energy Policy Act (EPACT) of 1992 beneath an oil derrick in Maurice, LA (see figure 4.2), declaring "a new era in which Government acts not as a master but as a partner and the servant" to the fossil fuel energy industry.[82] Many of the most climate-adverse elements of the 1991 National Energy Strategy were converted into law with the signing of the EPACT of 1992. The statute guaranteed free trade of natural gas, effectively mandated natural gas export to nations with whom the United States had a free-trade agreement, provided financial incentives to oil and gas drillers, committed funds to the research and development of new "clean coal" and natural gas technologies, and promised to export fossil fuel technologies to other nations.[83]

At a signing ceremony for the EPACT of 1992, President Bush said:

> Two years ago our administration proposed a national energy strategy. It was a blueprint to promote economic growth and make the country more secure. . . . But now our efforts have borne fruit, and this afternoon, right here, and it's fitting it happens in the shadow of a drilling rig, we're going to sign the Energy Policy Act of 1992.[84]

Finally, the first Bush administration, despite knowing that vehicles could get up to 50 mpg with then-current technology, and up to 80 mpg with

Figure 4.2
President Bush signs the Energy Policy Act of 1992, which assists the implementation of his National Energy Strategy. *Source*: US Department of Energy, "October 24, 1992: Energy Policy Act of 1992," https://www.energy.gov/management/october-24-1992-energy-policy-act-1992.

continued improvements in technology,[85] made only a modest improvement in the CAFE standards, raising them to the previous level of 27.5 mpg from 26 mpg. The requirements for light trucks, however, were actually lowered to 20 mpg.[86]

In sum, just four months after signing the framework convention on climate change and promising the world at the Rio Earth Summit "forceful action" on climate change and international cooperation, President Bush signed the Energy Policy Act of 1992 that would further a generation-long dependence on fossil fuels and growth in greenhouse gas emissions. This aptly captures the first Bush administration's position on climate change: Admit that it's happening on the one hand, but at the same time, cast doubt on the science, while supporting the fossil fuel industry and expanding fossil fuel development on the other hand.

5 THE WILLIAM J. CLINTON ADMINISTRATION (1993–2001)

The Clinton administration came into office in 1993 with a clear understanding of the dangers of climate change and with stated determination to act to meet that challenge. Vice President Al Gore had years before established himself as a national and international leader on the climate issue, and his 1992 book, *Earth in the Balance*, focused attention on the need for far-reaching climate action and became a bestseller. In a very pertinent passage in the book, Gore wrote: "We can believe in the future and work to achieve it and preserve it, or we can whirl blindly on, behaving as if one day there will be no children to inherit our legacy."[1] Moreover, the responsibility to move forward on the UNFCCC, give substance to its ultimate objective, and breathe life to its commitments had fallen to the Clinton administration.

President Clinton consistently acknowledged the threat and the necessity of taking action, starting just after taking office with his Earth Day address on April 21, 1993: "Unless we act now, we face a future . . . where our children's children will inherit a planet far less hospitable than the world in which we came of age."[2] During that address, he announced his pledge to cut emissions to 1990 levels by 2000.[3]

Members of the new administration also spoke of the need to address climate change, with Vice President Gore leading the way. Agencies such as EPA and DOI continued to issue reports consistent with the scientific consensus on climate change and anticipating severe impacts. The IPCC, meanwhile, issued its Second Assessment Report in 1995.[4] Undersecretary of State Timothy E. Wirth heralded the Second Assessment Report and said that "the world's scientists have reached the conclusion that the world's changing climatic conditions are more than the natural variability of weather. Human beings are altering the Earth's natural climate system."[5]

Though the United States participated importantly in the international negotiations leading to the signing of the Kyoto Protocol in 1998, and indeed Vice President Gore participated personally, Congress never ratified that treaty. The prospects for ratification were so bleak that the agreement was never sent to the Senate. The federal government, taken as a whole, effectively abandoned the treaty promises it had just made, thus pushing the growing burden of climate impacts onto the children who, in Vice President Gore's words, would inherit our "environmental legacy."[6] These actions thus presented a stark juxtaposition between the US commitment to an international climate convention steeped in the concrete promises just reviewed, and the default on those commitments in favor of the fossil fuel status quo. This historical moment embodies the reckless and knowing disregard not merely of knowledge that could prevent great harm but also of clear legal commitment to act on that knowledge. Meanwhile, carbon emissions rose steadily throughout the 1990s as fossil fuel growth and consumption continued during the Clinton years.

GOVERNMENT KNOWLEDGE

The Clinton Administration Understood Climate Science and Recognized the Growing Domestic and International Consensus as to the Causes and Projected Impacts of Climate Change

Knowledge of climate science continued to solidify and strengthen both within and outside the administration. The Clinton administration oversaw the release of multiple climate scientific assessments by federal agencies.[7] Several examples are worth mentioning. A 1993 EPA report noted that "the atmospheric concentration of CO_2 is increasing at the rate of about 1.8ppm/year (0.5%/year)," and that "the current concentration is higher than at any time in the last 160,000 years."[8] The 1993 EPA report further recognized that drastic reductions in anthropogenic greenhouse gas emissions were needed to stabilize atmospheric CO_2 concentrations, otherwise "CO_2 concentrations [of] (450 to 500 ppm) . . . will be reached within a reasonably short time (assuming current emissions rates)."[9]

A February 1993 State Department "Policy Decision Paper" recognized: "The best scientific evidence indicates that the continued increase in greenhouse gas concentrations will cause the global climate to change."[10] Importantly, the document recognized that, in the context of achieving the UNFCCC's "ultimate objective," *stabilizing* atmospheric greenhouse gas concentrations "would require dramatic (60 percent) reductions in current greenhouse gas emissions," and this could only be achieved within the United States through additional policy measures.[11]

In October 1993, Katie McGinty, from CEQ, circulated a memorandum on the president's Climate Action Plan. The memorandum included a suite of other documents; one was titled "State of Scientific Understanding of Climate Change."[12] This document outlined what climate scientists understood "very well" and "reasonably well" by that time, which was consistent with what prior administrations understood, as reviewed here.[13] President Clinton clearly grasped the gravity of climate change. In his October 1993 remarks at the unveiling of his Climate Action Plan, President Clinton stated that climate change "is a threat to our health, to our ecology, and to our economy."[14]

CEQ continued to issue clear warnings and acknowledgment of the threat. In its 1996 report (issued in 1997), CEQ noted: "The average global temperature is projected to rise 2 to 6 degrees over the next century, leading to increased flood and drought, rising sea levels, agricultural disruption, the spread of infectious disease and other health effects. The longer we wait to reduce our emissions, the more difficult the job, and the greater the risks."[15] In its 1997 annual report (issued in 1998) CEQ stated: "Within a span of just 100 years, the United States became the world's largest producer and consumer of fossil fuels," and, "since 1860, it is estimated that global CO_2 concentrations have increased from about 280 parts per million to about 360 parts per million today, or about 30 percent. Roughly half of that increase has occurred since 1970."[16]

Internationally, the Second IPCC Assessment report in 1995 warned of a worsening problem presented by continued atmospheric CO_2 buildup: "With the growth in atmospheric concentrations of greenhouse gases, interference with the climate system will grow in magnitude and the likelihood of adverse impacts from climate change that could be judged dangerous will become greater."[17]

In an October 22, 1997, speech to the National Geographic Society, President Clinton endorsed the IPCC's findings just noted, saying, "Average temperatures are rising. Glacial formations are receding." Clinton also stated: "Make no mistake, the problem is real. And if we do not change our course now, the consequences sooner or later will be destructive for America and for the world."[18]

Clinton Administration Officials Understood the Risks Climate Change Posed to the Nation's Youth and Future Generations, and Their Responsibility to Protect Future Generations From These Risks

As early as his April 1993 Earth Day speech, President Clinton issued a stark warning about the threat of climate change and said that "the bounty of nature is not ours to waste. It is a gift from God that we hold in trust for future generations. Preserving our heritage, enhancing it, and passing it along is a great purpose worthy of a great people."[19] Two years later, on Earth Day 1995, President Clinton remarked: "This continent is our home, and we must preserve it for our children, their children, and all generations beyond."[20]

President Clinton acknowledged the necessity of acting on climate for the sake of our children on multiple other occasions. In remarks to the Business Roundtable, on June 12, 1997, he said: "Let's find a way to preserve the environment, to meet our international responsibilities, to meet our responsibilities to our children, and grow the economy at the same time."[21] On July 24, 1997, the president again spoke eloquently at a climate change discussion at The White House:

> To me, we have to see this whole issue of climate change in terms of our deepest obligations to future generations. . . . It is obvious that we cannot fulfill our responsibilities to future generations unless we deal responsibly with the

> challenge of climate change. . . . If we fail to act, scientists expect that our seas will rise one to three feet, and thousands of square miles here in the United States, in Florida, Louisiana and other coastal areas will be flooded. Infectious diseases will spread to new regions. Severe heat waves will claim lives. Agriculture will suffer. Severe droughts and floods will be more common. These are the things that are reasonably predictable. . . . I believe the science demands that we face this challenge now. I'm positive that we owe it to our children.[22]

On October 6, 1997, at the White House Conference on Climate Change, President Clinton again referenced the younger generation: "We do not want the young people who sat on these steps today, for whom 33 years will also pass in the flash of an eye, to have to be burdened or to burden their children with our failure to act."[23]

Vice President Gore also often spoke of the peril that government refusal to change our nation's energy system posed for children/future generations. The Vice President remarked at the July 24, 1997, climate change discussion at the White House that "all of this gives rise to great concern that we are committing future generations to a planet that is altered in profound ways that can cause great harm to future generations."[24]

The Clinton administration's focus on protecting children from environmental harms is further reflected in a 1996 EPA report announcing a national agenda to protect children's health from environmental threats.[25] EPA's agenda was reinforced by a 1997 Executive Order (EO 13045), which set out that "each Federal agency: (a) shall make it a high priority to identify and assess environmental health risks and safety risks that may disproportionately affect children; and (b) shall ensure that its policies, programs, activities, and standards address disproportionate risks to children that result from environmental health risks or safety risks."[26]

On April 5, 1995, Timothy E. Wirth, then Undersecretary of State, spoke at the UNFCCC Conference of the Parties. He noted the scientific consensus on climate change and the risk it posed to future generations. "Every major peer-reviewed study has suggested that the most likely scenario is for a 3 to 8 degree F warming if carbon dioxide doubles from preindustrial levels. By increasing the concentration of greenhouse gases in the atmosphere at a rate unknown in all of human history, we are rolling the dice—gambling with our children's and grandchildren's future."[27]

The CEQ also issued a warning referencing our children. Its 1996 report states, "Our rising emissions of greenhouse gases have begun to affect the world's climate, and unless we take action to reduce them, our children and grandchildren will pay the price."[28] The Proceedings of the September 1995 Conference on Human Health and Global Climate Change, organized by the National Science and Technology Council and the Institute of Medicine/National Academy of Sciences, also contained several references to the particular vulnerability of children to climate change impacts, especially increasingly frequent and severe heat waves.[29]

The Clinton Administration Understood That Severe Climate Impacts Would Result From Unrestrained Greenhouse Gas Emissions

The Clinton administration was also well aware of the likely impacts associated with climate change, such as rising seas and coastal inundation, resource strains, spread of infectious diseases (particularly to children), biodiversity loss, increased conflict, and extreme heat and precipitation.

Undersecretary Timothy E. Wirth, speaking at the Conference of the Parties (COP) in April 1995, recognized, "The steady buildup of greenhouse gases in our atmosphere threatens to raise sea levels, change hydrological cycles and damage many of the world's ecosystems."[30] A year later on July 17, 1996, he again spoke at the Second Conference of the Parties held in Geneva, Switzerland: "Human health is at risk from projected increases in the spread of diseases like malaria, yellow fever and cholera; Food security is threatened in certain regions of the world; Water resources are expected to be increasingly stressed. . . . Coastal areas—where a large percentage of the global population lives—are at risk from sea level rise."[31]

Vice President Gore delivered remarks at the Conference on Human Health and Global Climate Change in September 1995, which was organized at the request of the White House. His remarks show a keen understanding of climate science, particularly the projected effects of atmospheric carbon loading on humans and particular sectors of society, including children:

> How will global warming affect us? There are clearly profound implications at the regional level for food security, water supplies, natural ecosystems, loss of land due to sea level rise, and human health. A temperature increase of 2 to 8 degrees Fahrenheit is projected to double heat-related deaths in New York City, and triple the number of deaths in Chicago, L.A. and Montreal. And an increase of 8 degrees Fahrenheit may be correlated with an increase in the heat/humidity index of 12 to 15 degrees. The very young, the elderly, and the poor will be the ones most at risk. So will those with chronic cardiovascular and respiratory diseases. . . . Changing temperatures and rainfall patterns are predicted to also increase the spread of infectious diseases. Insects that carry disease organisms may now move to areas that were once too cold for them to survive. These new breeding sites and higher temperatures may also speed reproduction. Diseases we had hoped were just a memory in this country are suddenly a renewed threat.[32]

Federal agencies during the Clinton administration were also in the process of developing increasingly sophisticated knowledge of the present and projected impacts to specific ecosystems. For example, in its April 1993 report concerning the impacts of climate change on forests, the EPA found:

> Changes in soil conditions due to loss of forest cover could slow forest reestablishment. Consequently, there could be a shift in area from forest to non-forest vegetation. Fire frequencies are likely to increase in the region given increased temperatures, unchanged precipitation and higher potential evapotranspiration. . . .
>
> The distribution and composition of forests in Washington and Oregon could change substantially. . . . In central Oregon, total forested area is projected to decrease by almost half under a 5°C warming.[33]

A 2000 US Forest Service report, "The Impact of Climate Change on America's Forests," further solidified knowledge among federal agencies of the particular vulnerabilities of the nation's forests to climate change.[34] The Forest Service report rather presciently projected "longer fire seasons and potentially more frequent and larger fires in all forest zones (even those that do not currently support fire)," and confirmed that "'improved forest management' appears to offer the most cost-effective means to sequester additional C [carbon] in forest ecosystems in the short term."[35]

Federal agencies were also increasingly aware of the risks of climate change-induced sea-level rise to coastal ecosystems. A June 1997 report by the US Geological Survey recognized that according to IPCC sea-level rise projections, "major portions of the coastal zone would be permanently flooded."[36] The USGS report also found that sea-level rise could increase the severity of storm surge and increase the existing vulnerability of coastal wetlands.[37]

The Department of Interior, meanwhile, recognized the risks climate change posed to public lands, as evidenced by its May 1997 report, "Climatic Change in the National Parks, Wildlife Refuges and Other Department of Interior Lands in the United States."[38] The report found that

> The climate that has helped shape the vegetation and wildlife and other characteristics of the DOI lands is expected to shift, while the boundaries of the natural protected areas (parks, refuges, wilderness areas and other DOI lands) remain fixed. . . . The basic vulnerability of DOI lands under these climatic conditions may be characterized by species that fail to migrate, fail to adapt, migrate to a less-protected environment, or are otherwise placed at a competitive disadvantage.[39]

Finally, federal agencies were gaining increasing knowledge of the projected regional differentiation in climate change impacts.[40]

GOVERNMENT ACTION

The Byrd–Hegel Resolution, the Fall of the Kyoto Protocol, and the Rise of Debilitating Partisanship

The Kyoto Protocol was the first major agreement aimed at implementing the UNFCCC and achieving its "ultimate objective" to "stabiliz[e] greenhouse gas concentrations in the atmosphere at a level that would prevent dangerous anthropogenic interference with the climate system."[41] It was signed by roughly 190 countries that are parties to the convention, and was adopted in Kyoto, Japan, in December 1997.[42] In pertinent part, the protocol provided that the developed country parties (what the protocol called "Annex I" parties), but not the developing countries, agreed to reduce their overall greenhouse gases by at least 5 percent below their 1990 levels by the 2008–2012 period.[43] The US target was 7 percent.[44]

As the protocol was being negotiated, coal-state Senator Robert Byrd and his colleague Senator Chuck Hegel introduced Senate Resolution 98 in June 1997, and it promptly passed 95–0 the following month.[45] The resolution expressed the sense of the Senate that no protocol was acceptable unless it included the developing countries and would not harm the US economy.

The senators understood perfectly that they had just thrown a spanner into the Kyoto works. Contemporaneous reporting identified that "without the participation of the United States, the leading source of waste industrial gases, the agreement would collapse."[46] The developing world was adamant that it was not ready to undertake binding obligations and that the rich countries, those responsible for most past and current emissions, must act first and foremost. The issue of their inclusion was one of the hottest ones at Kyoto.

In order to strengthen public understanding of the climate issue, and thus improve its political prospects, and to salvage a stalemated negotiation in Kyoto, the Clinton administration undertook two important initiatives. In October 1997, headlined by the White House Conference on Climate Change, the administration launched a wide-ranging and extensive public education and media campaign aimed at communicating the message that climate change was real and was a serious problem.[47] Then, in December Vice President Al Gore made a special trip to Kyoto to help salvage the negotiations.

Both actions underscore the seriousness of the Clinton administration's desire to move the issue forward, but in the end neither was successful in the US context. The Kyoto Protocol emerged from final negotiations with the terms just described, thus guaranteeing its easy defeat in the Senate. As a result, the Clinton administration never submitted it to the Senate for ratification and the United States never joined, thus doing serious damage to the overall international process.

Meanwhile, the public education effort took a strange turn. It appears from studies of the effects of that effort on public attitudes that it sharpened the partisan divide on climate, a divide that would only grow in the future and become more debilitating. For example, one sophisticated polling effort led by Stanford University found that the percentage of strong Democrats who believed that global warming would happen in the future held steady at about 75 percent, before and after the educational effort, but that the percentage of strong Republicans actually fell from 67 to 55 percent! When the issue was seen as being championed by Democrats, Republicans fled.[48] The climate issue has become steadily more partisan, and progress on climate is now stymied in part by an entrenched partisan gridlock. A huge partisan divide is one factor contributing to the inability of our political system to act meaningfully to confront the climate crisis.

The Clinton Administration Continued the United States' Commitment to a Nonrenewable, Fossil Fuel–Based Energy System

Despite the Clinton administration's acknowledgment of the climate issue and impacts, fossil fuels continued to be central to the nation's energy system under President Clinton. President Clinton's Climate Change Action Plan, released October 18, 1993, sought to decrease US greenhouse gas emissions to 1990 levels by the year 2000, but the plan was not integrated into energy policy or implemented with real teeth.[49] Overall, US GHG emissions (see figure 5.1) and fossil use (see figure 5.2) continued to increase during the 1990s.[50]

As figure 5.2 shows, there was an increase in consumption of energy from petroleum, natural gas, and coal, while the consumption of energy from renewable sources remained steady. While some of the graphic presentations reflect only modest changes in US energy use and emissions over time, it is important to remember the bathtub analogy: Even if the flow into the tub (atmosphere) is constant, the tub will fill up. If you want to stop the buildup, you have got to turn off the spigot! US GHG emissions have remained extremely high throughout the long period being examined, and have contributed greatly to "filling up the (atmospheric) tub." Estimates vary, but a common result in climate modeling today is that climate

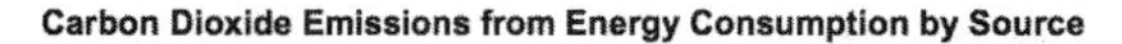

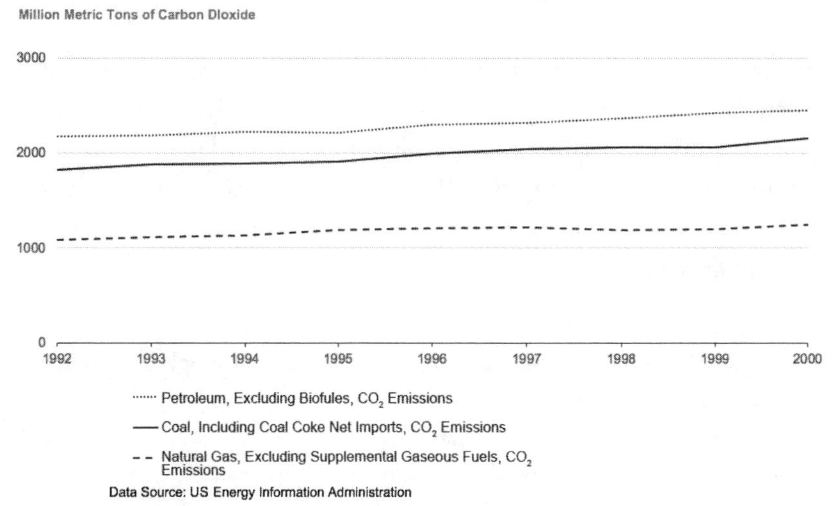

Figure 5.1

Carbon dioxide emissions from energy consumption by source. Data from US Energy Information Administration, *Monthly Energy Review August 2020*, 2020, table 11.1.

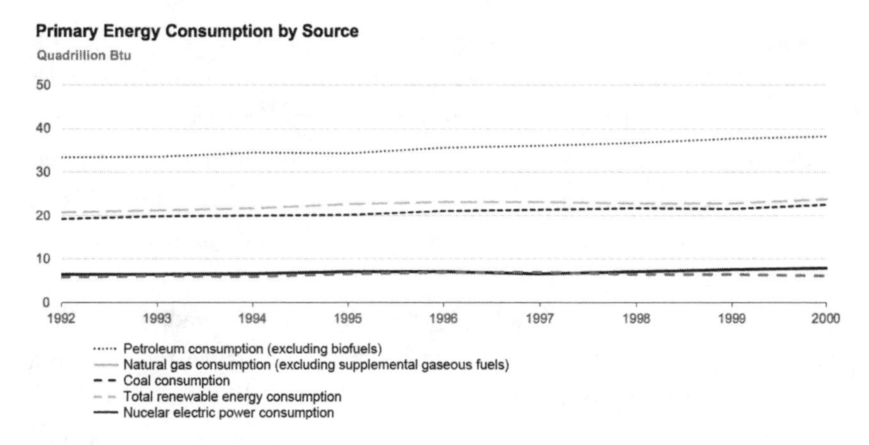

Figure 5.2

Primary energy consumption by source. Data from US Energy Information Administration, *Monthly Energy Review August 2020*, 2020, Table 1.3.

security requires that GHG emissions decline steadily to near zero by or before mid-century.

In addition to the Climate Action Plan's support for natural gas,[51] other federal agencies actively supported expanding fossil fuels. The Department of Energy, through its Office of Fossil Energy, continued to promote fracking and natural gas. An October 1999 report, "Environmental Benefits of Advanced Oil and Gas Exploration and Production Technology," touted the promise of natural gas drilling:

> Fortunately, in recognition of its environmental desirability, natural gas use has grown. Major advances in natural gas technology and supply have occurred over the past 25 years. New technology for finding, producing, transporting, storing, and using natural gas has been developed. Increasingly higher estimates of economically producible natural gas resources have improved the market's confidence in the reliability of long-term supplies.[52]

The Clinton administration directly supported the oil and gas industry in a number of ways, including by providing preferential tax treatment to the domestic oil and gas industry to encourage oil and gas production, providing federal funding for development of new technologies to increase oil extraction productivity, and providing continued support through research and development funding related to fossil fuel exploration to new places such as the Arctic and Alaska.[53] For example, the administration supported legislation to reduce federal royalty payments for offshore drilling.[54] While domestic oil production declined during the Clinton administration, oil imports increased significantly.

EPA had also proposed a "CO_2 cap with emissions trading" as "perhaps the most economically efficient mechanism possible for achieving emission reduction goals, irrespective of the level of the target."[55] Despite the fact that the EPA had received legal advice from the Office of General Counsel that CO_2 was an air pollutant under the Clean Air Act (CAA), and EPA had the authority to regulate CO_2 under the CAA, it declined to do so.[56] The Clinton administration also broke a campaign pledge in failing to increase the federal CAFE standard, with the administration arguing that "the impact of CAFE on GHG emissions before the year 2000 would be minimal" and that CAFE "isn't the silver bullet it's cracked up to be."[57]

The DOE, especially the DOE Office of Energy Efficiency and Renewable Energy, did do some important work during the Clinton administration to increase research and funding for renewable energy programs and implementation. For example, in 1997 DOE launched the Million Solar Roofs initiative, with the goal of installing solar energy systems on one million buildings in the United States by 2010.[58] (The one million goal was hit in 2016.[59]) DOE also launched the Wind Powering America initiative in 1999, with the goal of increasing the use of wind energy throughout the United States and providing 5 percent of the nation's energy with wind by 2020.[60] The DOE also adopted new appliance and equipment efficiency standards and allocated money for research and development of renewable energy supplies and the development of efficient vehicles. However, these efforts did not bend the fossil fuel emissions curve downward.

The Clinton administration fully acknowledged the climate problem, had abundant scientific evidence of the urgency and the dire consequences for children if action was not taken, and had policy options to bend the emissions trend downward. However, the national fossil fuel energy system and policies continued full throttle under the Clinton administration.

6 THE GEORGE W. BUSH ADMINISTRATION (2001–2009)

The administration of George W. Bush enacted policies favoring fossil fuels and failed to meaningfully address the worsening problem of climate change. Despite abundant scientific evidence and warnings, including two more assessment reports from the Intergovernmental Panel on Climate Change (IPCC), the second Bush administration continued to encourage and support the nation's reliance on fossil fuels. Elements in the administration, indeed in the Executive Office of the President, attempted to undermine public confidence in and understanding of climate science. According to a December 2007 report from the House Committee on Oversight and Government Reform, "the Bush Administration has engaged in a systematic effort to manipulate climate change science and mislead policymakers and the public about the dangers of global warming."[1] Publicly, President George W. Bush's White House did not deny the problem of global warming. It acknowledged the risks, especially during the early days of the administration. Yet the administration emphasized uncertainties in order to justify delaying action; it adopted what were at best half measures; it officially abandoned the Kyoto Protocol; and it committed the United States to increasing domestic fossil fuel production.

GOVERNMENT KNOWLEDGE

Climate Science and Climate Change

During the second Bush administration, research on carbon dioxide and other GHG emissions, as well as climate change generally, continued and climate models improved. The second Bush administration recognized climate change early on (though it also tried to emphasize uncertainties in the science). For example, in speaking at a G8 Environmental Ministerial Meeting in March 2001 in Trieste, Italy, new EPA Administrator Christine Todd Whitman remarked:

> Increasingly, there is little room for doubt that humans are affecting the Earth's climate, that the climate change we've seen during the past century is the result of human activity, and that we must continue our efforts to stop and reverse the growth in the emission of greenhouse gases.
>
> If we fail to take the steps necessary to address the very real concern of global climate change, we put our people, our economies, and our way of life at risk.[2]

In May 2001, the IPCC Third Assessment Report was released and reported new and even stronger evidence that global warming was caused by human activities.[3] The IPCC report predicted that global temperatures would increase by 1.4 to 5.8 degrees Celsius by 2100.[4] In June 2001, the National Research Council published a report in response to a request by the White House to identify uncertainties in climate science and to evaluate the findings and conclusions of the IPCC report.[5] The National Academy of Sciences (NAS) report affirmed the findings of the IPCC, and rather than casting doubt on climate change, stated unequivocally: "Greenhouse gases are accumulating in Earth's atmosphere as a result of human activities, causing surface air temperatures and subsurface ocean temperatures to rise. Temperatures are, in fact, rising."[6] The report predicted warming 3 degrees Celsius by 2100.[7]

A 2002 Department of State report, "U.S. Climate Action Report 2002," which was submitted to the UNFCCC Secretariat, reinforced the IPCC's findings. That report stated: "Greenhouse gases are accumulating in Earth's atmosphere as a result of human activities, causing global mean surface air temperature and subsurface ocean temperature to rise" and "Carbon dioxide from fossil fuel combustion was the dominant contributor. Emissions from this source category grew by 13 percent between 1990 and 1999."[8] The Department of State report also provided detailed evidence on the impacts of climate change, finding that (1) sea-level rise would lead to the loss of coastal wetland and put coastal communities at risk; (2) increases in heat waves were likely; (3) water shortages were likely to be exacerbated; (4) temperatures during the twenty-first century could rise 3–9°F; (5) the global sea level rose four to eight inches during the twentieth century at a rate significantly faster than the rate over the past several thousand years; and (6) climate models were predicting up to three feet of sea-level rise during the twenty-first century.[9] The Department of State report also found that the impacts of climate change would be worse in the twenty-first century than in the twentieth century, stating, "All climate model results suggest that warming during the 21st century across the country is very likely to be greater, that sea level and the heat index are going to rise more, and that precipitation is more likely to come in the heavier categories experienced in each region."[10] Finally, the Department of State report projected that US GHG emissions would rise by 43 percent between 2000 and 2020.[11]

Evidence of the human influence on climate and the potential impacts continued to accumulate during the George W. Bush era. The National Oceanic and Atmospheric Administration (NOAA), for example, released various climate studies and information related to tropical storms and hurricanes,[12] coral bleaching,[13] and the Arctic.[14] With respect to coral reefs, the study funded by NOAA warned that reefs were currently experiencing temperatures exceeding what the reefs have experienced in the last 400,000 years and that temperature increases of 1–2°C would be enough to wreak havoc and cause mass bleaching.[15]

In a hearing before the House Committee on Government Reform on July 20, 2006, the Director of NOAA's National Climatic Data Center, Thomas Karl, summarized the scientific knowledge of global climate change at that time:

> In conclusion, the state of the science continues to indicate that modern climate change is affected by human influences, primarily human induced changes in atmospheric composition. . . . Recent evidence suggests there will be changes in extremes of temperature and precipitation, decreases in seasonal

> and perennial snow and ice extent, sea level rise, and increases in hurricane intensity and related heavy and extreme precipitation.[16]

Also in 2006, NOAA published the first "State of the Arctic Report," finding that freshwater flow to the Arctic Ocean has been rising for the past several decades and is likely to continue to rise due to climate change.[17] The report also noted that "2002–2005 has been characterized by an *unprecedented* series of extreme ice extent minima."[18] A 2006 Congressional Budget Office (CBO) report noted the intergenerational harm issue, stating that "the combustion of fossil fuels may create external costs that are borne by society as a whole—particularly by future generations."[19] That CBO report also stated there is a

> possibility that greenhouse gases could build up to a critical level, or threshold, in the atmosphere and thus could trigger a rapid increase in damages. . . . In order to avoid passing such a threshold, it may be necessary to develop fundamentally new technologies that could provide a large share of the world's energy needs without releasing carbon or that could sequester similarly large shares of carbon emissions.[20]

The IPCC finalized its Fourth Assessment Report in 2007. The report contained a strong acknowledgment that humans are responsible for most of the observed planetary warming: "Most of the observed increase in global average temperatures since the mid-20th century is *very likely* due to the observed increase in anthropogenic GHG concentrations."[21] In a February 2, 2007, press release on the report, Dr. Sharon Hays, the leader of the US delegation at the meeting approving the report and Associate Director/Deputy Director for Science at the White House Office of Science and Technology Policy, described the IPCC report's significance:

> This Summary for Policymakers . . . will serve as a valuable source of information for policymakers.
>
> It reflects the sizeable and robust body of knowledge regarding the physical science of climate change, including the finding that the Earth is warming and that human activities have very likely caused most of the warming of the last 50 years.[22]

Later that year, President George W. Bush referenced the IPCC report in remarks during a meeting on energy security and climate change. "A report issued earlier this year by the U.N. Intergovernmental Panel on Climate Change concluded both that global temperatures are rising and that this is caused largely by human activities," the president said. "When we burn fossil fuels, we release greenhouse gases into the atmosphere, and the concentration of greenhouse gases has increased substantially."[23] This statement unequivocally demonstrates the president's understanding of the basic issue.

In 2007, the US Climate Change Science Program published a report, "Scenarios of Greenhouse Gas Emissions and Atmospheric Concentrations," that included various scenarios for CO_2 concentrations through the end of the twenty-first century, including multiple scenarios in which CO_2 concentrations were stabilized (i.e., no longer rising). While the lowest stabilization scenarios would have stabilized CO_2 concentrations at 450 ppm by 2100 (still too high), the report nonetheless illustrates that the federal government was evaluating and modeling various CO_2 concentration trajectories through the end of the 21st century.[24] The report noted that the stabilization scenarios "require a transformation of the global energy system" and that fossil fuel use and energy consumption must be reduced.[25]

The US Climate Change Science Program, in 2008, released a report describing their recent research and progress.[26] Among other important findings, the report noted that there have been abrupt changes in the global climate in the past over periods of a few years to decades and that rapid changes were already being observed.[27] For example, the report noted that there have been "dramatic declines in Arctic sea ice," which models were underestimating, and the Arctic could be seasonally ice-free earlier than IPCC projections.[28] Also in 2008, the EPA released a report by the US Climate Change Science Program, "Analyses of the Effects of Global Change on Human Health and Welfare and Human Systems," which described threats such as more powerful hurricanes and heat waves, decreased snowpack and shrinking water supplies in the West, and spread of disease.[29] A *Washington Post* story on the report noted, "The EPA report yesterday was less notable for its warnings . . . than for its source. The Bush administration has resisted the conclusion that increasing temperatures will harm human health, but in yesterday's report, that finding was unmistakable."[30]

Prominent government climate scientists like Dr. James Hansen, meanwhile, continued to issue important warnings. An article published in *Reuters* in 2007 quoted Dr. Hansen, then director of NASA's Goddard Institute for Space Studies, stating, "We're a lot closer to climate tipping points than we thought we were. . . . If we are to have any chance in avoiding the points of no return, we're going to have to make some changes."[31]

The following year, twenty years after his landmark testimony to Congress, Dr. Hansen and nine colleagues (Dr. Hansen and two other authors were working for the federal government at the time for the Goddard Institute for Space Studies) published a very important scientific paper indicating that stabilizing the climate system requires limiting atmospheric CO_2 concentrations to a maximum of 350 parts per million—a concentration that had already been exceeded. The article found that

> If humanity wishes to preserve a planet similar to that on which civilization developed and to which life on Earth is adapted, paleoclimate evidence and ongoing climate change suggest that *CO_2 will need to be reduced from its current 385 ppm to at most 350 ppm.*[32]

President George W. Bush acknowledged the need to reduce GHG emissions, stating in 2001, "Our approach [to reduce greenhouse gas emissions] must be consistent with the long-term goal of stabilizing greenhouse gas concentrations in the atmosphere."[33] However, the second Bush administration did not take a position on what CO_2 concentration would prevent dangerous interference with the climate. As James L. Connaughton, Chairman of the CEQ under President George W. Bush, put it:

> The President has reaffirmed America's commitment to the goal of stabilizing atmospheric greenhouse gas concentrations at a level that will prevent dangerous interference with the climate. At the same time, the President noted that given current scientific uncertainties, no one knows what that level is.[34]

When Connaughton was asked again in 2007 for the White House's position on "dangerous climate change," Connaughton replied, "We don't have a view on that."[35]

RENEWABLES AND EFFICIENCY

As with previous presidential administrations, the federal government during the second Bush administration was aware of renewable energy sources that could reduce reliance on fossil fuels and alleviate the threats posed by climate change.

The 2002 Department of State report, "U.S. Climate Action Report 2002," contained a whole chapter on policies and measures that would reduce GHG emissions.[36] The Department of State report considered specific ways to reduce GHG emissions from various sectors of the economy, including electricity, transportation, industry, buildings, agriculture and forestry, and the federal government. More specifically, the report considered ways to incentivize energy efficiency in the residential and commercial sector; ways to reduce waste and GHG emissions from industry; ways the federal government could support wind, solar, hydropower, and other renewable energy sources; and ways to reduce GHG emissions from the transportation sector, including through the use of hydrogen as a fuel source.[37]

A 2004 General Accounting Office (GAO) report found that the Midwest has enough wind power potential "to meet a significant portion of the nation's electricity needs."[38] The GAO report also noted that "most of the nation's wind potential remains untapped. Wind power's growth will depend largely on the continued availability of federal and state financial incentives, including tax credits, and expected increases in prices for fossil fuels."[39] In a 2005 presentation, Dr. Harlan Watson, the Senior Climate Negotiator and Special Representative for the Department of State, set forth various federal programs that could be used to reduce GHG emissions, including fuel economy standards, energy efficiency standards, renewable energy tax incentives, hybrid/fuel cell vehicle tax incentives, clean air rules, and biological sequestration.[40] In 2006, a National Renewable Energy Laboratory conference paper found that "the potential for wind to supply a significant quantity of energy in the United States is enormous," and that 2,000–4,000 million metric tons of carbon could be avoided by 2030 through the use of wind energy.[41]

A Congressional Budget Office report noted how different government policies can encourage the use of fossil fuels, such as federal funds for highway construction or tax provisions to promote oil and gas production, while other policies can discourage the use of fossil fuels, such as a federal gas tax and subsidies for mass transit.[42] It also observed that setting a price or cap on carbon would be a way to correct the market failure due to the negative externalities of carbon pollution and climate change.[43] As noted in a 2007 GAO report:

> It is unlikely that DOE's current level of R&D funding or the nation's current energy policies will be sufficient to deploy advanced energy technologies in the next 25 years. Without sustained high energy prices or concerted, high-profile federal government leadership, U.S. consumers are unlikely to change their energy-use patterns, and the United States will continue to rely upon its current energy portfolio.[44]

The 2007 GAO report also noted that the proportion of US energy coming from renewable sources, 6 percent, had not increased since 1973.[45]

While the second Bush administration was aware of renewable energy sources, the administration focused its research and development on fossil fuels and not renewable energy research. As one GAO report from 2008 stated:

> DOE has focused its R&D on increasing domestic production primarily by improving exploration technologies, extending the life of current oil reservoirs, developing drilling technology to tap into deep oil deposits, and addressing environmental protection. DOE officials stated that if the oil R&D program continues, it would focus on such areas as enhanced oil recovery technologies and expanding production from independent producers.[46]

As noted in a 2006 GAO report, DOE's budget for renewable energy had been cut by over 85 percent, from 1975 to 2005 (in real terms), despite the need to advance renewable energies.[47] The administration was also aware that one of the hurdles to more widespread adoption of renewable energy sources was the low cost of fossil fuels.[48]

GOVERNMENT ACTION

While the robust body of scientific knowledge showed that human-caused climate change was a clear and present danger, the second Bush administration took actions that were at odds with the science. Energy lobbyists with access to the White House pushed administration officials to play up the uncertainties and play down the risks. The result was a misrepresentation of the strength of climate science by the administration and a clear attempt to undermine the findings.

As for actual climate policies, President George W. Bush offered only modest initiatives as he relinquished US climate leadership in the international arena, and his EPA refused to regulate carbon pollution. All the while, the administration encouraged domestic fossil fuel production and oversaw an energy policy favorable to the interests of the fossil industries.

Misleading the Public on Scientific Findings

Evidence of political interference with scientific information as well as inappropriate influence from the oil industry appeared early in the second Bush administration. A March 2002 document faxed to CEQ's Philip Cooney from ExxonMobil's Randy Randol revealed Exxon's perspective on the state of climate science and suggested focusing more on the uncertainties. Exxon claimed that scientific knowledge of climate change was limited and suggested research be geared towards areas of uncertainty. It also criticized the IPCC for its "all-too-apparent bias . . . to downplay the significance of scientific uncertainty and gaps."[49]

Cooney, a former lobbyist for the American Petroleum Institute who would later go on to work for ExxonMobil, was in direct communication with the fossil fuel industry and its associated think tanks while at CEQ. He received emails[50] and memos from Myron Ebell at the Competitive Enterprise Institute, for example. One such email, dated June 3, 2002, suggested, "Our only leverage to push you in the right direction is to drive a wedge between the President and those in the Administration who think that they are serving the president's best interests by pushing this rubbish,"[51] with rubbish referring to the "U.S. Climate Action Report 2002" prepared for the UNFCCC.

Cooney was later found to have played a role in manipulating scientific reports. According to a House of Representatives Committee on Oversight and Government Reform report, "CEQ Chief of Staff Phil Cooney and other CEQ officials made at least 294 edits to the administration's *Strategic Plan of the Climate Change Science Program* to exaggerate or emphasize scientific uncertainties or to deemphasize or diminish the importance of the human role in global warming."[52] The House report stated:

> The Committee's 16-month investigation reveals a systematic White House effort to censor climate scientists by controlling their access to the press and editing testimony to Congress. The White House was particularly active in stifling discussions of the link between increased hurricane intensity and global warming. The White House also sought to minimize the significance and certainty of climate change by extensively editing government climate change reports. Other actions taken by the White House involved editing EPA legal opinions and op-eds on climate change.[53]

The House Committee came to "one inescapable conclusion: the second Bush Administration has engaged in a systematic effort to manipulate climate change science and mislead policymakers and the public about the dangers of global warming."[54] When Cooney's line-by-line edits became public, the ensuing scandal was enough to force his resignation.[55]

An internal investigation also revealed climate science censoring by NASA's Public Affairs Office during the second Bush administration. According to a June 2008 NASA Office of Inspector General investigative summary:

> Officials in the NASA Headquarters Office of Public Affairs did, in fact, manage the release of information concerning climate change in a manner that reduced, marginalized, and mischaracterized the scientific information within the particular media over which that office had control. Further, on at least one occasion, the Headquarters Office of Public Affairs denied media access to a NASA scientist, Dr. Hansen, due, in part, to that office's concern that Dr. Hansen would not limit his statements to science but would, instead, entertain a policy discussion on the issue of climate change.[56]

WEAK CLIMATE POLICY

By the turn of the new century, it was difficult (but as we shall see, not impossible!) for any administration to simply walk away from the climate issue. The second Bush administration addressed the issue, but not meaningfully and often harmfully.

Political interferences aside, President George W. Bush at least superficially accepted climate science and announced measures in response. "My administration is committed to a leadership role on the issue of climate change. We recognize our responsibility, and will meet it—at home, in our hemisphere, and in the world," the president said in 2001.[57] One approach the administration took, rather than to take steps to actually reduce GHG emissions and fossil fuels use, was to instead emphasize the need for more research. Consistent with this approach, in 2002, the interagency Climate Change Science Program (CCSP) was established to coordinate and direct US research efforts in the area of climate change.[58] The research was to be conducted as part of the US Global Change Research Program (GCRP) and Climate Change Research Initiative (CCRI). Despite creation of the CCSP, it was never given enough funding to fully carry out its research agenda and programs. As an NRC report explained, "The present CCSP budget does not appear to be capable of supporting all of the activities in the strategic plan. . . . There is no evidence in the plan or elsewhere of a commitment to provide the necessary funds for these newer or expanded program elements."[59]

The president also made a pledge to reduce greenhouse gas emissions relative to economic activity—otherwise known as reducing GHG intensity. Specifically, the president called for a GHG intensity reduction of 18 percent over ten years. Speaking in Silver Spring, Maryland, on February 14, 2002, he said:

> My administration is committed to cutting our Nation's greenhouse gas intensity . . . by 18 percent over the next 10 years. This will set America on a path to slow the growth of our greenhouse gas emissions and, as science justifies, to stop and then reverse the growth of emissions.[60]

It should be noted that reducing GHG intensity—the amount of GHG emitted per dollar of gross domestic product (GDP)—which President Bush claimed "science justifies," doesn't actually reduce total emissions unless intensity is declining faster than the economy is growing. President George W. Bush's proposal called for an intensity decline of less than 2 percent a year, so that even if achieved (it wasn't), his proposal was unlikely to stabilize, much less reduce, actual US GHG emissions.

President George W. Bush also approved research and development (R&D) for climate change science and technology. According to an open letter written by Bush science advisor John Marburger and CEQ Chair James Connaughton in February 2007, President Bush "committed nearly $3 billion annually—more than any other country in the world—to climate change technology research and deployment programs" from 2003 to 2006.[61] This was spent on energy efficiency technology programs, consumer information campaigns, and incentives, but also on research into sequestration of carbon dioxide, which was planned to be sold abroad by US coal companies, and adaptation measures.[62]

President George W. Bush's EPA started to work on a multipollutant cap-and-trade initiative called Clear Skies. At one point early in the legislative process, carbon dioxide was considered for inclusion.[63] Ultimately, and even though the eventual bill excluded carbon dioxide as a pollutant, the Clear Skies legislation never passed Congress.

Furthermore, the White House declared it would veto climate legislation such as the Lieberman–Warner Climate Security Act[64] and moved to block an EPA proposal for regulating carbon emissions from mobile sources.[65] The second Bush administration also opposed a national renewable energy portfolio standard that would have increased renewable energy, and it opposed increasing the CAFE standards (by regulation or statute), which remained at 27.5 mpg throughout the second Bush presidency.[66]

The second Bush administration also reversed a Clinton administration determination that GHGs could be regulated under the Clean Air Act, taking the position that GHGs were not air pollutants.[67] However, after the Supreme Court issued its decision in *Massachusetts v. EPA*, the president directed the EPA, Department of Transportation (DOT), and other federal agencies to "take the first steps toward regulations that would cut gasoline consumption and greenhouse gases."[68] In a letter to the president, EPA Administrator Stephen L. Johnson wrote:

> The Supreme Court's *Massachusetts v EPA* decision still requires a response. That case combined with the latest science of climate change requires the Agency to propose a positive endangerment finding, as was agreed to at the Cabinet-level meeting in November. . . . The state of the latest climate change science does not permit a negative finding, nor does it permit a credible finding that we need to wait for more research.[69]

Despite commitments by President George W. Bush and pressure from the EPA, the White House Office of Management and Budget moved to block an EPA report outlining how the EPA could regulate CO_2 from mobile and stationary sources.[70] By the time President George W. Bush left office, the EPA had not yet proposed a rule to regulate GHG emissions.[71]

On the international stage, President George W. Bush officially backed out of the Kyoto Protocol, already a dead letter in the United States, claiming the agreement would hurt our economic growth and jobs. "The Kyoto Protocol would have required the United States to drastically reduce greenhouse gas emissions. The impact of this agreement, however, would have been to limit our economic growth and to shift American jobs to other countries," the president remarked in April 2008.[72] However, even though the United States withdrew from the Kyoto Protocol, the second Bush administration still made efforts to undermine international climate negotiations. For example, the United States firmly opposed proposed binding international GHG emission reduction targets that called for reductions in GHG reductions of 25–40 percent below 1990 levels by 2020. In an effort to undermine the international negotiations on the binding targets, the United States advanced parallel international talks for voluntary goals.[73]

THE BUSH ADMINISTRATION EMBRACES FOSSIL FUELS

The second Bush administration understood and embraced the prominence of the federal government in setting national energy policy. Unfortunately, the administration's national policies embraced fossil fuels, not renewable energy and efficiency. Soon after taking office, President George W. Bush signed Executive Order 13211, which stated, "The Federal Government can significantly affect the supply, distribution, and use of energy."[74] The EO required all federal agencies to prepare a statement explaining how their regulatory authority could have "any adverse effects on energy supply, distribution, or use."[75] A memorandum to the heads of executive departments regarding the executive order made clear that the purpose of the executive order was to eliminate regulatory barriers to fossil fuel production.[76] Another executive order, signed on the same day, called for federal agencies to expedite their review of energy-related projects.[77]

Also during the first few months of his presidency, Bush created the National Energy Policy Development Group, headed by Vice President Dick Cheney, and gave it the mission to "develop a national energy policy designed to help the private sector, and as necessary and appropriate Federal, State, and local governments."[78] In May 2001, the National Energy Policy Development Group released a report with over 100 recommendations to the president. The National Energy Policy report did mention renewable energy but mostly emphasized enhancing fossil fuel production. It stated:

> A primary goal of the National Energy Policy is to add supply from diverse sources. This means domestic oil, gas, and coal. It also means hydropower and nuclear power. And it means making greater use of non-hydro renewable sources now available. . . . Currently, the U.S. has enough coal to last for another 250 years.[79]

The National Energy Policy report recommended that federal agencies incentivize offshore oil and gas development, reduce oil and gas royalties, boost natural gas production, and expand natural gas pipelines, among other actions to reduce regulatory hurdles to expanding fossil fuels extraction and transportation. It concluded that US oil consumption would increase by 33 percent and demand for electricity would increase by 45 percent in the next twenty years.

The recommendations in the National Energy Policy report were largely influenced by industry groups and petroleum lobbyists. The *Washington Post* reported that in the year of its existence the task force had met with about 300 groups and individuals.[80] The policy developed for and executed by the Bush administration included encouraging oil and gas production, investments in oil and gas infrastructure, and increased exploration and production of oil, natural gas, and coal.[81] The vast majority of the groups the task force consulted with represented oil and gas interests or fossil fuel companies and trade groups, including the National Mining Association, the Interstate Natural Gas Association of America, and the American Petroleum Institute.

President Bush highlighted oil and gas during remarks announcing his energy plan in Minnesota on May 17, 2001, stating: "New technology makes drilling for oil far more productive. . . . My administration's energy plan anticipates that most new electric plants will be fueled by the cleanest of all fossil fuels, natural gas. . . . I will call on Congress to pass legislation to bring more gas to market."[82]

The president also touted coal, the dirtiest and most emissions-intensive fossil fuel (though he claimed it could be burned cleanly). "Increasing our energy security begins with a firm commitment to America's most abundant energy sources—source, and that is coal. . . . Clean coal technology advances—will advance, and when it does, our society will be better off."[83]

The second Bush administration's favoritism toward fossil fuels was evident in the Energy Policy Act of 2005. This legislation contained an exemption for hydraulic fracturing from the Safe Drinking Water Act, a gift to the oil and gas industry from Vice President Dick Cheney, a former Halliburton CEO, that was dubbed the "Halliburton Loophole."[84]

The second Bush administration used the advent of the Energy Policy Act of 2005 to advocate for opening up Arctic National Wildlife Refuge (ANWR) to oil and gas development. Although the final legislation did not include drilling in ANWR, it is notable that the ANWR was not the only protected area under federal control that the administration wished to open to fossil fuel exploitation. President George W. Bush supported "increasing the production of traditional energy resources on the Outer

Continental Shelf (OCS), Federal onshore lands, and Indian lands, consistent with the National Energy Policy."[85] A September 2005 memorandum to the president further demonstrates the administration's pro-fossil-fuel agenda: "Your advisors are developing options to: 1. increase refining capacity; 2. address natural gas shortages, in both the short and long term; and 3. increase oil and natural gas production."[86]

This push to increase oil and gas production came during a crucial time when America could have been leading the effort to address climate change. Instead, the second Bush administration, heavily influenced by the oil and gas industry, largely abandoned that responsibility. The concept of using less energy was openly scoffed at by Vice President Cheney, who famously said in a 2001 speech: "Conservation may be a sign of personal virtue, but it is not a sufficient basis for a sound, comprehensive energy policy. . . . The aim here is efficiency, not austerity."[87] While the administration did pay lip service to renewables and energy efficiency, policies and funding measures showed the real priorities: securing more oil supplies, developing natural gas resources, and producing as much coal as possible. Its priorities were reflected in its federal leasing, permitting, subsidies, and infrastructure in support of the fossil fuel energy system. As a result, greenhouse gas emissions increased throughout the second Bush administration until the Great Recession of 2008.

Notwithstanding the emerging impacts of climate change, the ever-growing literature on climate science, and the availability of renewable energy sources to provide energy for the nation, the Bush administration strategy was to cast doubt on the science and focus on the need for more research while prioritizing short-term economic interest, especially those of the fossil fuel companies. The administration opposed any binding measures to reduce GHG emissions and instead called for weak, voluntary measures, both nationally and internationally. Unfortunately for the climate system and the youth plaintiffs in this case, the administration was successful in expanding fossil fuel development and production.

7 THE BARACK OBAMA ADMINISTRATION (2009–2017)

President Barack Obama appeared to take the threat of climate change seriously and did more than any other president to address it. By this point the unequivocally robust climate science and already evident impacts were impossible to ignore. The federal government, from EPA and research institutions to the Department of Defense, fully and openly acknowledged the scientific warnings.

In 2014, the Department of Defense acknowledged climate change will affect its ability to defend the nation and poses immediate risks to US national security. The report stated: "Among the future trends that will impact our national security is climate change. Rising global temperatures, changing precipitation patterns, climbing sea levels, and more extreme weather events will intensify the challenges of global instability, hunger, poverty, and conflict. They will likely lead to food and water shortages, pandemic disease, disputes over refugees and resources, and destruction by natural disasters in regions across the globe."[1]

Similarly, in 2016, the National Intelligence Council reported that climate change is likely to pose wide-ranging national security challenges for the United States over the next 20 years through threats to the stability of other countries.[2]

Notably, EPA issued in 2009, after years of litigation and an eventual Supreme Court decision mandating that it do so,[3] an Endangerment Finding for greenhouse gases based on the overwhelming scientific information on climate change and its impacts.

Reports continued to add to the body of knowledge underpinning the danger of anthropogenic climate change.

President Obama, like many of his predecessors, spoke of the problem—particularly the risk it posed to future generations—and did take some steps toward significant action, particularly during his second term in office. He announced a Climate Action Plan, and his EPA issued regulations addressing emissions from motor vehicles and power plants. President Obama also participated in international climate summits, culminating in the Paris Agreement in late 2015.

But by simultaneously pursuing an "all-of-the-above" national energy policy and approving more fracked gas and unconventional fossil fuel pipelines than any president before him, his administration ensured that fossil fuels remained an entrenched engine of our economy and the dominant fuel of America's energy system. Under the Obama administration, domestic oil and gas production increased, as did fossil fuel exports. This locked in further greenhouse gas warming at a time when the climate science alarm sirens were blaring. Despite President Obama's concerns and piecemeal efforts to take action on climate, the problem only worsened during his administration.

GOVERNMENT KNOWLEDGE

Knowledge of Climate Science, Impacts, and Children

The state of scientific understanding of human-caused climate change was well advanced by the time Barack Obama took office as the forty-fourth president, and this knowledge continued to advance during his eight years in office.

Early on in his first term, in June 2009, the US Global Change Research Program released its Second National Assessment detailing climate change impacts in the United States. The report explained the expected impacts across a variety of areas, including the public health and social costs. As the report noted about population shifts and development choices, "Climate change is likely to exacerbate these challenges as changes in temperature, precipitation, sea levels, and extreme weather events increasingly affect homes, communities, water supplies, land resources, transportation, urban infrastructure, and regional characteristics that people have come to value and depend on."[4]

Statements made in US Senate committees during the summer of 2009, meanwhile, confirmed that climate change represented a risk to national security[5] and to the national park system.[6]

Also in 2009, the US Geological Survey (USGS) published a report warning about how rising ocean temperatures, ocean acidification, and sea-level rise were threatening coral reefs. The USGS report stated that "business-as-usual scenarios that project increasing carbon dioxide concentrations in the atmosphere and oceans from burning fossil fuels and deforestation portend unprecedented changes in the distribution, abundance, and survival of coral communities and life in the global oceans."[7]

By the end of 2009, the EPA issued its most sweeping declaration of the threat of climate change, officially publishing an Endangerment Finding under Section 202(a) of the Clean Air Act.[8] EPA stated in its finding that "warming of the climate system is unequivocal" and that "the evidence provides compelling support for finding that greenhouse gas air pollution endangers the public welfare of both current and future generations. The risk and the severity of adverse impacts on public welfare are expected to increase over time."[9] In an April 28, 2010, congressional subcommittee hearing, EPA Administrator Lisa Jackson pointed out the scientific consensus supporting this finding:

> I found in December 2009 that motor vehicle greenhouse gas emissions do endanger Americans' health and welfare. I am not alone in reaching that conclusion. Scientists of the 13 Federal agencies that make up the U.S. Global Change Research Program have reported that unchecked greenhouse gas emissions pose significant risk to the well-being of the American public. The National Academy of Sciences has stated that the climate is changing . . . and that those changes will transform the environmental conditions on Earth unless countermeasures are taken.[10]

Following EPA's groundbreaking Endangerment Finding, multiple reports were issued under the Obama administration that corroborated the already strong scientific knowledge on the danger of global warming.[11]

In 2011, a National Research Council report looked at the impacts of climate change based on varying levels of CO_2 concentration and temperature rise. The NRC report found that under a "business as usual" scenario a CO_2 concentration of 1000 ppm could be reached by 2100, and stabilizing the atmospheric CO_2 concentration at any target level "would require reductions in total emissions of at least 80 percent (relative to any peak emission level)."[12] A graphic in the report illustrates the cumulative carbon emissions and temperature change for various CO_2 concentration trajectories (see figure 7.1). Importantly, "the report concludes that certain levels of warming associated with carbon dioxide emissions could lock Earth and many future generations of humans into very large impacts."[13]

Additionally, a 2011 Congressional Research Service (CRS) report discussed using a "science-based approach" to evaluate a "safe" or "tolerable" amount of climate change.[14] This science-based approach focuses on the relationship between temperature change, projected climate change impacts, and GHG concentrations. While the CRS report did not take a position on what CO_2 concentration is "safe," it is still noteworthy that the CRS was considering various CO_2 concentrations (including 350 ppm) and the corresponding temperature increase.

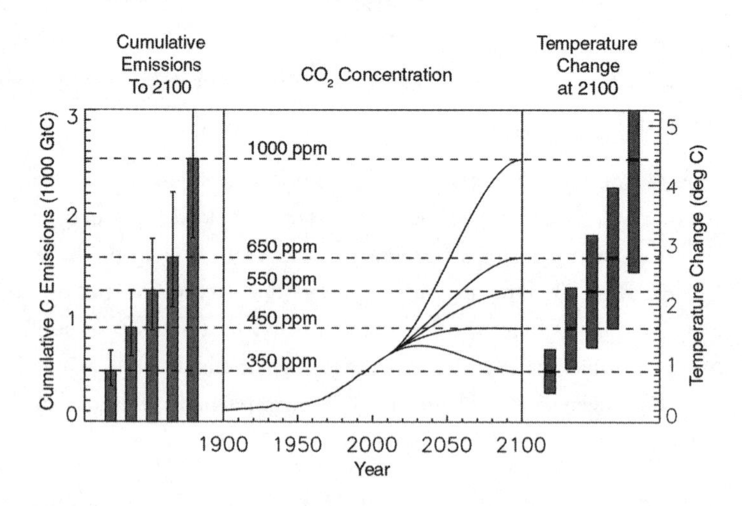

Figure 7.1
CO_2 concentration scenarios reaching between 350 and 1,000 ppm by 2100. *Source:* National Research Council, *Climate Stabilization Targets*, 104.

In 2011, NOAA's Office of Coast Survey retired thirty-five place names for islands, bays, passes, and ponds that had ceased to exist because of sea-level rise.[15]

The EPA issued the second edition of its "Climate Change Indicators in the United States" report in 2012.[16] Among other findings, the 2012 EPA report noted that the extent of Arctic sea ice has been decreasing over time, and that September 2012 was the smallest level ever recorded.[17] Two years later the third edition report of "Climate Change Indicators" came out, finding climate change contributes to an increase in heat-related deaths, as well an increase in Lyme disease, allergies, and asthma-related problems. The 2014 report also found that climate change threatens human health, and in particular the health of children, by exacerbating the risk of wildfires, hurricanes, the loss of coastal land, the loss of glacial water storage, and the loss of ice in the Arctic.[18] The EPA also issued a report in 2013 specifically about children, in which it found that children are especially vulnerable to the impacts of climate change such as higher temperatures and heat waves, increased air pollution, and spreading infectious diseases.[19]

The US Global Change Research Program continued to be a leading authority on climate science, issuing two additional reports in 2014 and 2016. The 2014 Third National Climate Assessment report notably stated: "Climate change, once considered an issue for a distant future, has moved firmly into the present."[20] According to findings of the Third National Climate Assessment, "The global warming of the past 50 years is primarily due to human activities, predominantly the burning of fossil fuels. . . . U.S. average temperature has increased by 1.3°F to 1.9°F since record keeping began in 1895; most of this increase has occurred since about 1970."[21] The 2016 assessment detailed the health impacts of climate change and noted disproportional impacts on classes of people like children, those with preexisting medical conditions, communities of color, indigenous peoples, and coastal populations.[22] The executive summary of that report bluntly stated: "Climate change threatens human health and well-being in the United States."[23]

In the Inventory of US Greenhouse Gas Emissions and Sinks 1990–2016, EPA reported that in 2016 fossil fuel combustion accounted for 93.5

percent of all CO_2 emissions in the United States.[24] In 2017 CO_2 emissions from energy consumption in the United States were approximately 5,153 million metric tons, with 29 percent from natural gas, 26 percent from coal, and 45 percent from petroleum.[25]

Another key climate report that came out during Obama's second term was the IPCC's Fifth Assessment Report (AR5). The language in that report was the strongest yet in describing the observed warming and associated impacts. The report stated: "Human influence on the climate system is clear, and recent anthropogenic emissions of greenhouse gases are the highest in history. . . . Warming of the climate system is unequivocal. . . . Continued emission of greenhouse gases will cause further warming and long-lasting changes in all components of the climate system, increasing the likelihood of severe, pervasive and irreversible impacts for people and ecosystems."[26]

President Obama clearly understood the severity of these scientific warnings and in particular the impacts on children. During his Second Inaugural Address in January 2013 he made an important reference to children and future generations in the context of climate change. "We will respond to the threat of climate change, knowing that the failure to do so would betray our children and future generations," the president remarked.[27] He again spoke of the threat during his 2013 State of the Union Address:

> Now, it's true that no single event makes a trend. But the fact is the 12 hottest years on record have all come in the last 15. Heat waves, droughts, wildfires, floods—all are now more frequent and more intense. We can choose to believe that Superstorm Sandy, and the most severe drought in decades, and the worst wildfires some states have ever seen were all just a freak coincidence. Or we can choose to believe in the overwhelming judgment of science—and act before it's too late.[28]

In a candid interview with the *New York Times* published in September 2016, President Obama revealed his reaction to his briefings on the latest climate information. "'My top science adviser, John Holdren, periodically will issue some chart or report or graph in the morning meetings,' he said, 'and they're terrifying.'"[29]

Knowledge Regarding Renewable Energy Sources

In 2009, only 4 percent of the US electric power industry was made up of nonhydro renewable energy sources.[30] The Obama administration, however, was aware of the federal government's role in shaping national energy policy, and knew of viable renewable energy sources available to reduce dependence on fossil fuels.

In 2010, the DOE projected that 20 percent of the nation's electrical supply could come from wind energy by 2030.[31] In 2011, the DOE, after noting that more than 80 percent of total US primary energy and over 95 percent of US transportation fuel comes from fossil resources, said:

> The Department has a unique role in defining end-use standards for appliances and other electronic devices and informs the vehicle fuel economy standards set by the Department of Transportation. . . . The United States has the opportunity to lead the world in a new industrial revolution to manufacture the clean energy technologies we need and create the jobs of the future. To ensure America's competitiveness in this century and achieve our energy goals, we must develop and deploy clean energy technologies in our nation.[32]

Then, in 2012, the DOE, through its National Renewable Energy Laboratory, found that by 2050, "Electricity supply and demand can be balanced in every hour of the year in each region with nearly 80% electricity from renewable resources, including nearly 50% from variable renewable generation."[33]

Speaking in 2013, President Obama said:

> The path towards sustainable energy sources will be long and sometimes difficult. But America cannot resist this transition, we must lead it. We cannot cede to other nations the technology that will power new jobs and new industries, we must claim its promise. That's how we will maintain our economic vitality and our national treasure.[34]

In 2017, the DOE, in discussing the nation's energy system, stated that "the Federal Government will play a major role in managing the challenges and taking advantage of the opportunities that the 21st-century grid presents."[35]

Finally, the Obama administration knew that any delay in reducing emissions would significantly increase the eventual cost of the necessary reductions. A report from the Executive Office of the President found that for each decade of delay, mitigation costs increased by approximately 40 percent. According to the White House report, "each year of delay means more CO_2 emissions, so it becomes increasingly difficult, or even infeasible, to hit a climate target that is likely to yield only moderate temperature increases."[36]

GOVERNMENT ACTION

Compared to the previous administration, the Obama administration went further in terms of actions addressing the climate problem. However, President Obama did not reverse decades of government support for fossil fuel energy, and he continued to support fossil fuels in many ways, embracing domestic oil and gas production and touting an "all of the above" energy strategy that implied indefinite reliance on fossil fuels.

Climate Change Policy Under Obama

Obama's climate policy consisted of trying to implement policies to cut carbon pollution, limit other greenhouse gas pollutants, encourage a clean energy economy, improve federal climate adaptation efforts, and return the United States to a position of leadership in the global climate negotiations.[37] I highlight the administration's most notable efforts here.

First, following the Supreme Court's ruling in *Massachusetts v. EPA* in 2007 that the EPA had the authority and duty to regulate greenhouse gas emissions under the Clean Air Act, Obama's EPA promulgated new vehicle emission and mileage standards starting in 2010 for light-duty vehicles for model years 2012 through 2016 raising vehicle standards to an average of 30.2 miles per gallon (mpg), up from 27.5 mpg before the rule.[38] In October 2012, standards for model years 2017–2025 were added, which would require manufacturers to produce light-duty vehicles for model years 2017–2025 to achieve industry-average fuel efficiency equivalents of 54.5 mpg.[39] A rule for the EPA and the National Highway Traffic Safety

Administration also later established standards for medium- and heavy-duty trucks.[40]

The cornerstone of Obama's climate policy was his Climate Action Plan, announced on June 25, 2013. Speaking on a hot summer day at Georgetown University and pausing to wipe sweat from his forehead,[41] President Obama revealed his plan to bypass Congress and direct EPA to set limits on carbon pollution:

> This is a challenge that does not pause for partisan gridlock. It demands our attention now. And this is my plan to meet it—a plan to cut carbon pollution, a plan to protect our country from the impacts of climate change; and a plan to lead the world in a coordinated assault on a changing climate. . . . Today, for the sake of our children, and the health and safety of all Americans, I'm directing the Environmental Protection Agency to put an end to the limitless dumping of carbon pollution from our power plants and complete new pollution standards for both new and existing power plants.[42]

The EPA followed through and issued the first-ever power plant carbon pollution standards, starting with new power plants in September 2013 and existing power plants in August 2015. The resulting regulations were called the Clean Power Plan (CPP).[43] The CPP was the centerpiece of America's climate policy under Obama. When the United States officially pledged to reduce carbon emissions by 26 percent to 28 percent below 2005 levels by 2025—its "intended nationally determined contribution" offered ahead of the Paris climate talks in 2015[44]—the CPP was advertised as central to that goal.

However, the defendants stated in their Answer to Youth Plaintiffs' First Amended Complaint that "the Clean Power Plan is not intended to 'preserve a habitable climate system.'"[45] (Moreover, it was never implemented, and the Trump administration is replacing it with a new rule that will not require significant changes to the energy industry or reductions in GHG emissions.[46]) In the Obama administration's report to the UNFCCC on its progress to meet its stated goals, the defendants showed that even if the CPP were fully implemented, US emissions would essentially decrease, and then continue a slight downward trend or possibly increase,

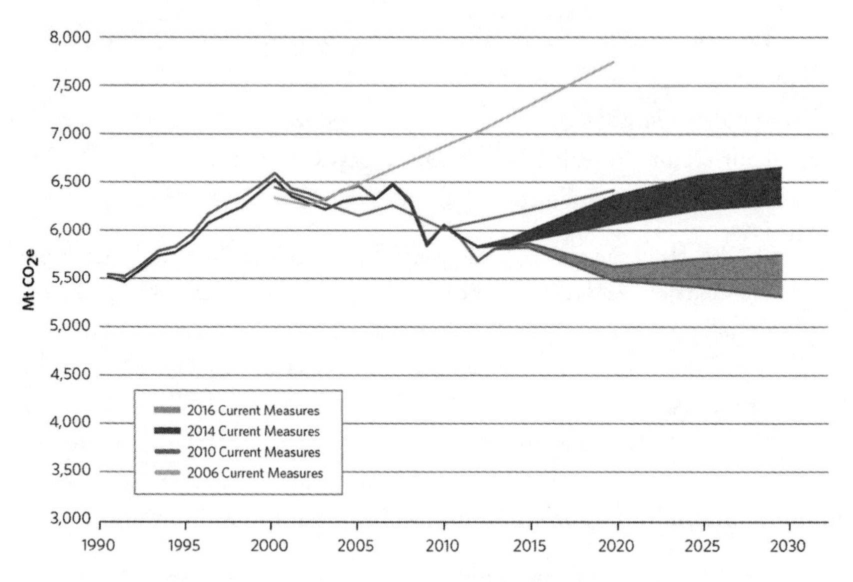

Figure 7.2

Source: US Department of State, *Second Biennial Report of the United States Under the United Nations Framework Convention on Climate Change*, 2016, 35. Ex. E-261.

but not decrease at the rate urgently needed to respond to the crisis (figure 7.2). The defendant DOE's EIA also confirmed a brief decline in emissions followed by a flat line with CPP implementation, as illustrated in figure 7.3.

The defendant DOE under Obama also finalized energy efficiency rules, including those governing large-scale commercial air conditioners, heat pumps, and furnaces, which could have avoided significant carbon emissions just through efficiencies had President Trump later not directed the halt to implementation of those rules.

The Obama administration also promulgated new standards to lower methane emissions from oil and gas development and landfills and recommitted to reducing hydrofluorocarbons (HFCs) and other greenhouse gases.

The initial adoption of the Paris Agreement in December 2015 was hailed as a landmark moment in international climate policy. The United

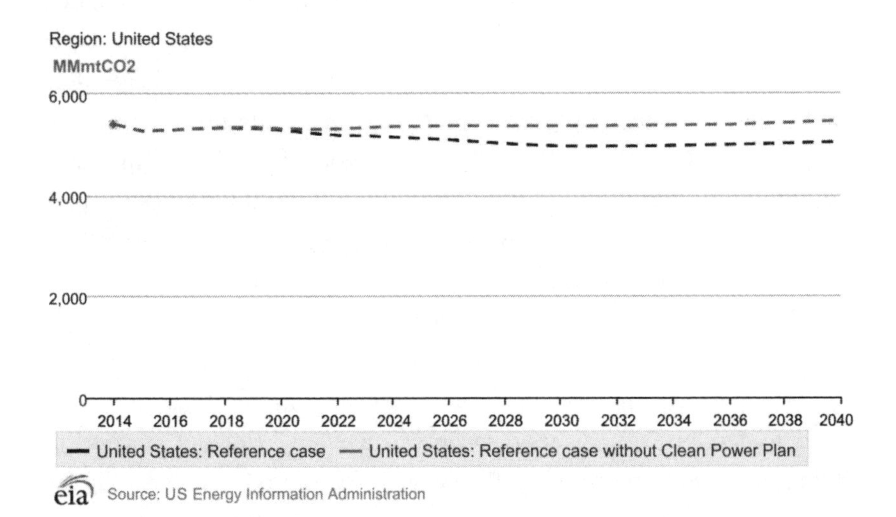

Figure 7.3
Energy use and related statistics: Carbon dioxide emissions projected with and without Clean Power Plan implementation through 2040. Data from US Energy Information Administration, Energy Use & Related Statistics: Carbon Dioxide Emissions, https://www.eia.gov (graph created through EIA's interactive table browser using EIA data from the publication "Annual Energy Outlook 2016" and the table "Energy Consumption by Sector and Source" comparing the "Reference Case" with the "Reference Case without Clean Power Plan").

States under the Obama administration signed onto the Paris Agreement, which declared a global goal of "holding the increase in the global average temperature to well below 2°C above pre-industrial levels and pursuing efforts to limit the temperature increase to 1.5°C above pre-industrial levels."[47] This built upon the 2009 Copenhagen Accord that recognized the need to limit global temperature rise below 2°C.[48] Despite the majority of negotiating parties supporting the adoption of a 1.5°C target in the treaty's text, US negotiators advocated—ultimately unsuccessfully—for a 2°C target in the treaty's preamble only.[49] The United States did, however, succeed in having the language in the key sections of the Paris Agreement regarding climate change mitigation and finance altered so that the sections only constituted aspirational or voluntary provisions, rather than binding obligations under international law.[50]

Around the time of the Paris Agreement, the Obama administration also released its Mid-Century Strategy for Deep Decarbonization, focused on cost-effective decarbonization pathways.[51] That plan was a start of a national plan but was never implemented and has since been shelved like all previous plans and roadmaps considered by prior administrations.

President Obama also led the way to greater federal investments in clean energy and approved the first-ever large-scale renewable energy project on federal public lands. He also announced a 21st Century Clean Transportation Plan to scale up national electric vehicle infrastructure.[52]

President Obama also signed several executive orders pertaining to climate change and climate adaptation. An executive order issued October 5, 2009, directed the new Interagency Climate Change Adaptation Task Force to develop "approaches through which the policies and practices of the agencies can be made compatible with and reinforce" a national climate change adaptation strategy.[53] This task force was replaced with an interagency Council on Climate Preparedness and Resilience (established through a 2013 executive order on preparing the United States for the impacts of climate change).[54] These actions represent an attempt at a coordinated federal response to prepare for the inevitable effects of a changing climate.

However, the Obama administration's climate policy efforts fell far short of a reasonable response to the dangerous climate situation facing young people, especially when viewed in light of the defendants' overall energy policies.

Obama Administration Energy Policy

Recognizing that climate change and energy policy were closely intertwined, the Department of Energy cautioned in its May 2011 Strategic Plan that a "business as usual" energy strategy "will imperil future generations with dangerous and unacceptable economic, social, and environmental risks."[55] DOE further noted: "As part of prudent risk management, our responsibility to future generations is to eliminate most of our carbon emissions and transition to a sustainable energy future. . . . To ensure America's competitiveness in this century and achieve our energy goals, we must develop and deploy clean energy technologies in our nation."[56]

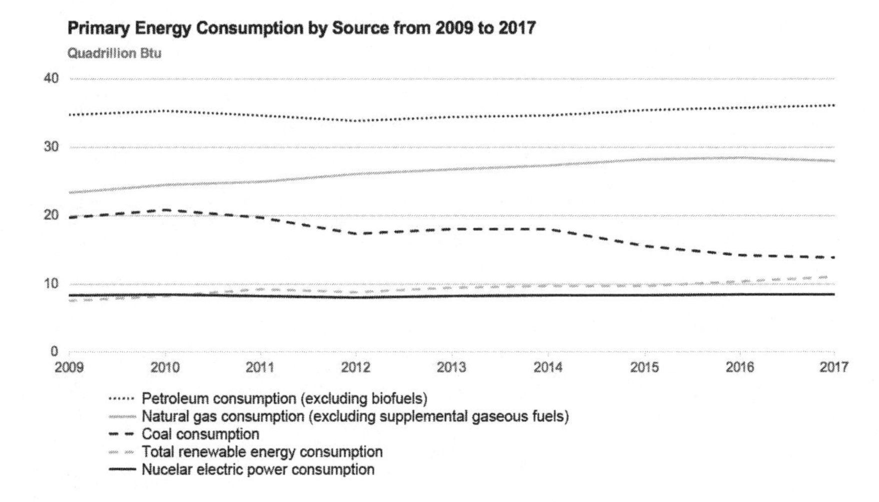

Figure 7.4

Primary energy consumption by source from 2009 to 2017. Data from US Energy Information Administration, *Monthly Energy Review August 2020*, 2020, table 1.3.

Throughout the Obama years, however, fossil fuel supplied the vast majority of our energy. Both natural gas and petroleum consumption increased during the Obama administration while coal consumption decreased (see figures 7.4 and 7.5). The national energy policies of the administration did not align with the urgent need to act on climate or the recommendations of defendant agencies, including the DOE.

A July 2011 CBO report, "The Effects of Renewable or Clean Electricity Standards," pointed out the dominant role that coal and natural gas played in the electricity portfolio. "Currently, only about 10 percent of U.S. electricity is produced from renewable sources of energy," the report noted. "The bulk of electricity is produced using coal (45 percent), natural gas (24 percent), and nuclear power (19 percent)."[57] A report from the Executive Office of the President noted that 40 percent of coal production in the United States came from federal lands and that the regulations and administrative process governing leasing of federal coal have been largely unchanged since the late 1970s and early 1980s.[58]

A 2017 DOE report stated:

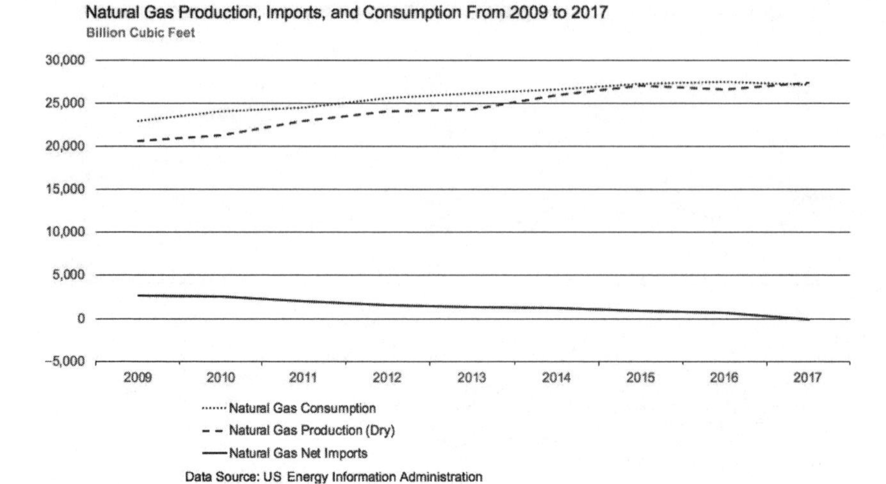

Figure 7.5

Natural gas production, imports, and consumption from 2009 to 2017. Data from US Energy Information Administration, *Monthly Energy Review August 2020*, 2020, table 4.1 (preliminary data for 2019).

> A sustained, 40-year Federal policy commitment has enabled a robust, global oil market; a diversity of petroleum suppliers; the world's largest strategic oil reserve; international mechanisms for concerted action in the event of disruptions; increased domestic oil production; a shift away from oil-fired power generation; more-efficient vehicles; and a host of other benefits.[59]

The Obama administration especially embraced oil and natural gas. According to a 2014 Congressional Research Service report, "In August 2014, approximately 4.1 million barrels per day (bbl/d) of petroleum products, NGLs, and other liquids were exported from the United States—up from an average of nearly 1.4 million bbl/d in 2007."[60] According to that CRS report, between 1970 and 2008 US crude oil production was steadily declining, but since 2009, crude oil production was rapidly increasing and expected to continue to rise.[61]

President Obama himself spoke about exploiting the country's supply of oil and natural gas. In his 2012 State of the Union Address he said:

> Over the last three years, we've opened millions of new acres for oil and gas exploration, and tonight, I'm directing my administration to open more than

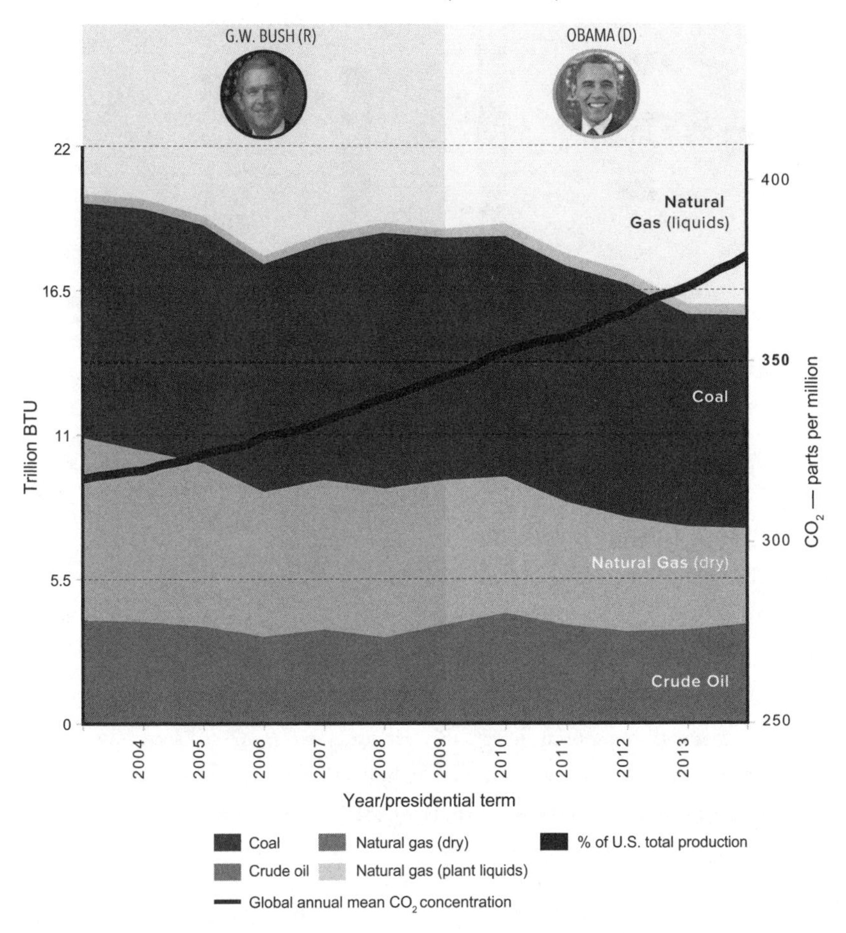

Figure 7.6

US fossil fuel extraction from federal public lands during the G. W. Bush and Obama administrations. Data from US Energy Information Administration, *Sales of Fossil Fuels Produced from Federal and Indian Lands, FY 2003 through FY 2014*, Washington, DC, July 2015, https://www.eia.gov/analysis/requests/federallands/pdf/eia-federallandsales.pdf; National Oceanic and Atmospheric Administration Earth System Research Laboratory, accessed July 2018, https://www.esrl.noaa.gov/gmd/ccgg/trends/data.html.

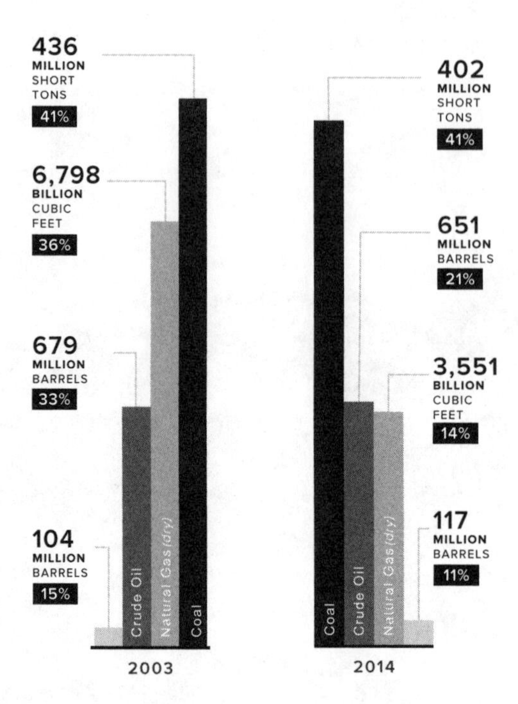

Figure 7.6 (continued)

> 75 percent of our potential offshore oil and gas resources. Right now—right now—American oil production is the highest that it's been in eight years. . . . We have a supply of natural gas that can last America nearly 100 years. And my administration will take every possible action to safely develop this energy.[62]

A few months later, speaking at TransCanada Pipe Yard near Cushing, Oklahoma, President Obama bragged about domestic petroleum production under his administration, stating:

> Over the last three years, I've directed my administration to open up millions of acres for gas and oil exploration across 23 different states. We're opening up more than 75 percent of our potential oil resources offshore. We've quadrupled the number of operating rigs to a record high. We've added enough new oil and gas pipeline to encircle the Earth and then some. So we are drilling all over the place—right now. That's not the challenge. That's not the problem. In fact, the problem . . . is that we're actually producing so much oil and

gas . . . that we don't have enough pipeline capacity to transport all of it where it needs to go.[63]

In his 2013 State of the Union, the president described his energy policy as "all of the above," indicating that the United States would not be reducing its reliance on fossil fuels any time soon. "Now, in the meantime, the natural gas boom has led to cleaner power and greater energy independence. We need to encourage that. And that's why my administration will keep cutting red tape and speeding up new oil and gas permits," Obama remarked. "That's got to be part of an all-of-the-above plan."[64] President Obama's "all-of-the-above" plan included significant levels of fossil fuel extraction continuing on federal public lands, as illustrated in figure 7.6.

Under President Obama's leadership, the United States began to take steps, primarily through EPA, toward regulating CO_2 and reengaging in the global arena through international negotiations on climate treaties. However, nearly all of this progress was undermined by efforts elsewhere in the Obama administration to expand US fossil fuel exploitation and production—and would be reversed altogether by the next president, Donald Trump.

8 THE DONALD J. TRUMP ADMINISTRATION (2017–2020)

As I indicated in my overview to this expert report, my purpose here is "to provide expert testimony regarding the historic knowledge of the US federal government (including defendants) of climate change, climate science, and alternative pathways to power the nation's energy system other than fossil fuels," as well as "the decisions made by the US federal government to devise and pursue energy policies and, in particular, to maintain a fossil fuel–based energy system." Thus far, this expert report supports the conclusion that the defendants continued from the Carter years through to the Obama administration to plan for, support, invest in, permit, and otherwise foster a national fossil fuel–based energy system.

Now we come to President Donald J. Trump, inaugurated January 20, 2017. The vast accumulation of knowledge about the climate issue, its origins and solutions, from the Johnson administration forward can leave no doubt at this point that the defendants knew all they needed to know to take the issue with utmost seriousness. Further, the defendants' own scientific reports strongly recommended pathways to transition away from fossil fuels, including through conservation, efficiency, and solar and other renewables. Notwithstanding this information, the Trump administration has acted with reckless disregard of the lives of young people and future generations, wielding the powers of the presidency and federal agencies to roll back (modest) climate protections then in place or proposed, disparage and undermine climate science, and abandon international climate agreements and US engagement internationally, all the while carrying to a new extreme the defendants' historic support for fossil fuel extraction and an energy economy based largely on fossil fuels. Indeed, the Trump administration has sought to eliminate virtually every climate protecting initiative of prior administrations, as inadequate as they on balance have been. The president has even discussed invoking the Defense Production Act to marshal support for coal use.[1]

As I was preparing this rewrite of developments during the Trump administration, I received a periodic and authoritative overview of recent developments in climate science and impacts.[2] The report focused on specific developments found in recent months:

- Following a record-breaking heatwave, unprecedented fires have raged across Siberia, burning an area larger than Greece.
- Arctic sea ice extent was the lowest in the satellite record, its volume in September 2018 three times lower than that of September 1979.
- Death Valley in California reached the hottest temperature ever recorded, 130 degrees F.
- Global emissions of methane, a greenhouse gas twenty-eight times more powerful than CO_2, reached their highest level on record.
- Flooding in Europe in recent decades has reached its highest levels in 500 years.
- Since the 1950s, the frequency, duration, and severity of heat waves have accelerated over many world regions.
- Polar bears are projected to become nearly extinct by the end of this century.
- Beavers are now colonizing Arctic tundra areas in Alaska and Canada.
- Projections estimate a large increase of yellow fever deaths by mid-century, as well as a major expansion in the zika range.

The list goes on, and this periodical addressed only a modest share of recent developments. It is powerfully suggestive, though. In covering several decades in my expert report, I have quoted leading authorities on the range of ill effects climate change could bring. It is plain to anyone who follows the news that a myriad of long-predicted effects are now here every day, affecting human lives and natural ecosystems and limiting future choices. We no longer have to expect the climate crisis. We are experiencing it.

No administration in modern times has acted with such outward disregard for and disdain of climate science or proceeded to promote policies that so imperil the well-being of today's children and future generations. Even if many of Trump's efforts to roll back past gains and prevent others prove to fail in the end for one reason or another, precious time and momentum will have been squandered—and indeed squandered at a critical moment in the history of the issue when developments seem to be coming together to offer honest hope. Meanwhile, despite some recent stirrings and legislative proposals since 2018, Congress has continued its failure to legislate substantive, science-based solutions throughout the Trump years, defaulting in its leadership to the Executive Branch.

I have followed the climate issue closely for forty years now. I have never been as frightened for my grandchildren as I am now, both because of the late hour to try to stop the onslaught of dangers and because of the disturbing disregard of this administration for science and the rights of young people and all future generations.

GOVERNMENT KNOWLEDGE

Knowledge of Climate Science and Impacts

Given the information available to prior administrations, there is no question that, as of the Trump administration, the defendants knew the science behind climate change and knew emissions from continued burning of high levels of fossil fuels were causing the climate crisis. Yet, as discussed in the following, the Trump administration took the deprivation of America's children and posterity to new extremes.

Redressing harms and protecting Americans from the climate crisis means following the science, a science that the Trump administration has gone beyond ignoring, to expressing outright hostility. One example of Trump's dangerous disregard for government scientific reports is his handling of the Fourth National Climate Assessment (NCA4). The periodic National Climate Assessment (NCA) is prepared on a regular basis by the

US Global Change Research Program (USGCRP), which comprises 13 federal agencies. The NCA provides a comprehensive assessment of the current understanding of climate change science.[3] The first NCA was released in 2000. Between 2002 and 2009, USGCRP produced 21 assessment products and then in 2009 released NCA2. The third report (NCA3) was published in 2014.

Volume I of NCA4, released in 2017, stated that many temperature and precipitation extremes have become more frequent, more intense, or longer in duration.[4] It noted that climate models are consistent with these trends continuing, which may make natural disasters more frequent or more intense. Further, Volume I of NCA4 reported that some climate change effects, including sea-level rise and increased coastal flooding, could lead to the dispersal of pollutants, posing a risk to public health.

Volume II of NCA4 was released on November 23, 2018, the day after Thanksgiving. This release date was seen by many as an attempt to downplay its findings by releasing them on a day when fewer Americans than usual were paying attention to the news.[5]

At over 1,500-pages, Volume II of NCA4 is a massive report, emphasizing the dire threat that human-caused global warming poses to the United States and its citizens.[6] "Earth's climate is now changing faster than at any point in the history of modern civilization, primarily as a result of human activities," researchers state in Volume II. The NCA4 details the climate and economic impacts US residents will see if drastic action is not taken to address climate change. "The impacts of global climate change are already being felt in the United States and are projected to intensify in the future," the researchers wrote. The last few years have smashed records for damaging weather in the United States, costing nearly $400 billion since 2015. In a worst-case scenario, the researchers project that climate change could deliver over $500 billion a year in damages by the end of the century. The NCA4 concludes that climate change threatens the health and well-being of the American people by causing increasing extreme weather, changes to air quality, the spread of new diseases by insects and pests, and changes to the availability of food and water. It observes that these climate-related impacts will only get worse and their costs will mount dramatically if carbon emissions continue unabated.

The NCA4 is but one example during the Trump years of the federal government recognizing that, due to climate change, US residents are forced to cope with dangerously high temperatures, rising seas, deadly wildfires, torrential rainfalls, and devastating hurricanes. Yet the Trump administration has persistently ignored these types of warnings from scientists and economists. Trump himself discredited the NCA4's findings that climate change damage to US lives, livelihoods, and property could cost many hundreds of billions of dollars each year, saying, "I don't believe it."[7]

The same day Volume II of NCA4 was released, the US Geological Survey (USGS) released a report estimating the amounts of life-cycle emissions from oil, gas, and coal development on federal public lands.[8] This USGS report looked at data from 2005–2014; it did not include data from the Trump administration, which has pushed even further to expand fossil fuel development on public lands. The USGS report showed that GHG emissions from federal energy production on public lands are a significant source of total US emissions. The USGS report concluded that emissions from fossil fuels produced on federal lands represent, on average, 24 percent of national emissions for CO_2, 7 percent for CH_4, and 1.5 percent for N_2O over the 10 years included in this estimate. The USGS report shows that emissions from fossil fuels produced on federal lands (and waters) in twenty-eight states and two offshore areas in 2014 were 1,279 million metric tons of CO_2.[9] For 2014, emissions from fossil fuels from these federal lands (and waters) represented 23 percent of national CO_2 emissions.[10]

Attack on Science and Analysis, and Ethics

Trump and his top administration officials have repeatedly misled the public about the basics of climate change. As recently as 2018, Trump said, "I don't know that it's manmade," and disavowed the need to act.[11] Other top administration officials conceded that humans play a role in climate change but questioned how much.[12] Their statements conflict with scientific studies by the federal government, such as NCA4, which concluded climate change, pollution, and other human activities (like deforestation) are the only factors that can account for the global warming occurring over the last century. Trump and his top officials have also disputed scientific findings about the potentially catastrophic impacts of climate change. Scott Pruitt, Trump's first EPA Administrator, even questioned whether climate change "necessarily is a bad thing."[13] Officials from the White House's Office of Legislative Affairs, the Office of Management and Budget, and the National Security Council barred the Office of the Geographer and Global Issues, a State Department intelligence agency, from submitting written testimony to the House Intelligence Committee warning climate change is "possibly catastrophic."[14] The Department of the Interior's top climate change scientist was reassigned to an accounting role, despite no training in accounting, after he highlighted the dangers climate change poses for Alaska's Native communities.[15] EPA officials blocked agency scientists and contractors from presenting research about climate change and related ecological issues at a professional conference.[16] Agency officials at the Departments of Agriculture, Energy, and the Interior have similarly prevented staff from attending scientific conferences.[17]

In 2018, EPA staff members were given misleading talking points to use with the public.[18] Rather than acknowledge the broad scientific consensus about human-caused climate change, the talking points focused on the lack of perfect precision in measuring humans' impacts and on the "continuing debate" about human causation that the Trump administration has promoted. The Trump administration even altered or removed mention of climate change on numerous webpages of agencies across the US government, and some federal climate change information resources are no longer being maintained. At the EPA, for example, a website devoted to climate change information was removed from the EPA's active website and relegated to an archival site where the information is no longer updated.[19] References to climate change were virtually eliminated from a DOI webpage.[20]

During his run for president, Trump called for dismantling the EPA.[21] During Trump's first eighteen months in office, the EPA lost on net more than 1,200 employees, or 8 percent of its workforce. Buyouts, attrition, and a government-wide three-month hiring freeze left the EPA's workforce below its 1988 level. In those eighteen months, nearly 1,600 workers left the EPA, while fewer than 400 were hired.

While Congress has prevented Trump from completely carrying out his pledge, his administration has been relentless in its efforts to roll back public health and environmental protections, weaken enforcement of those protections, and cripple the EPA's capacity to address new and existing problems.

Fossil fuel interests provided substantial funds for both Trump's election and his reelection efforts.[22] Trump has stocked federal agencies with industry lobbyists, lawyers, and other representatives.[23] Several have already been forced out after ethical problems were exposed. For example, appointed Trump's first EPA Administrator, Scott Pruitt was Attorney General of Oklahoma, where he spearheaded lawsuits against EPA rules regulating pollution from fossil fuel and other industries. After eighteen months as EPA Administrator, Pruitt resigned in the midst of thirteen federal investigations into allegations of legal and ethical violations, including spending abuses, first-class travel, and use of EPA staff for personal tasks.[24] Trump's second EPA Administrator, Andrew Wheeler, was a lawyer and lobbyist with a firm that represented several polluting industries, including coal company Murray Energy. Murray's chief executive officer (CEO) handed the Trump administration a road map for rolling back regulations on the coal industry. Key Senate-confirmed positions have been left unfilled.[25]

Meanwhile, Trump has appointed industry insiders to deputy positions wielding much of the power of the unfilled top positions. In May 2019, the President proposed a climate-science panel chaired by William Happer, a physicist on the National Security Council, who once said "the demonization of carbon dioxide is just like the demonization of the poor Jews under Hitler" and that the earth is suffering through a "CO_2 famine."[26] The Trump administration's response to the coronavirus shows what happens when federal agencies' missions are not respected and their capacity to identify and respond to crises is compromised.

The EPA's strategic plan for FY2018–FY2022 does not mention climate change.[27] The EPA's previous strategic plans discussed reducing climate pollution and adapting to climate change. Reference to climate change was also dropped from the strategic plan of the Federal Emergency Management Agency (FEMA).[28] Finally, in 2015, the US Government Accountability Office (GAO) recommended that the Executive Office of the President designate a federal entity to develop and periodically update a set of authoritative climate change observations and projections for use in federal decision making and create a national climate information system with defined roles for federal agencies and nonfederal entities.[29] The Executive Office of the President failed to agree or disagree with the GAO's recommendations and has not implemented the recommendations.[30]

The Trump administration has made sweeping changes to environmental impact statements (EISs) that would effectively reduce or even eliminate analysis of the climate change impacts of proposed major federal actions.[31] Federal agencies are required to prepare an EIS to help agencies, with public input, minimize environmental impacts of big projects like pipelines. More broadly, the Trump administration has limited scientific and economic assessments in ways that result in federal agencies deemphasizing relevant scientific evidence and undervaluing pollution control benefits. For example, it has changed how damage from climate change is estimated by ignoring much of the damage and slashing the value of future generations.[32] By low-balling estimates, the changes create a bias against taking action to combat and prepare for climate change. These estimates are used in cost–benefit analyses that influence federal decision making.[33]

The Trump administration has directed agencies to consider only climate change damage that occurs within the United States and to ignore climate change damage to the rest of the world. And in estimating climate change damage, the Trump administration estimated the social cost of carbon to be lower than prior estimates by making the calculation using a higher range of discount rates—3 to 7 percent—rather than 2.5 to 5 percent used for the prior estimates, which results in lower present values. This significant change to the "discount rates" greatly affects how the government values the lives, health, and welfare of future generations. As a result, the current federal estimates for domestic climate damages are about 7 times lower than the prior federal estimates (when both prior and current estimates are expressed in 2018 US dollars and calculated using a 3 percent discount rate).[34] As a mathematical matter, the highest rate used in the Trump administration approach makes the lives of unborn generations nearly worthless in its cost–benefit analyses.

Permitting, leasing, and authorizing fossil fuel exploration, development, extraction, and infrastructure results in "carbon lock-in," ensuring continued CO_2 emissions from such activities and infrastructure.[35] Simply stated, the Trump administration's policy of promoting construction of new fossil fuel infrastructure as well as fossil fuel leasing of federal resources will make it more difficult and costly to reverse an already dangerous warming.[36]

Impacts on Children

In 2017, the EPA proposed a two-year delay on implementation of a 2016 Obama-era rule designed to reduce climate-warming methane emissions, eliminating a requirement that oil and gas companies monitor and limit methane leaks from wells, compressor stations, and other operations. Obama's methane rule would have required oil and gas companies to detect and repair leaks of methane and other air pollution at drilling wells.

When issued, the agency's proposed rule stated, in relevant part, "*the EPA believes that the environmental health or safety risk addressed by this action may have a disproportionate effect on children.* . . . However, because this action merely proposes to delay the 2016 Rule, this action will not change any impacts of the 2016 Rule after the stay. Any impacts on children's health caused by the delay in the rule will be limited, because the length of the proposed stay is limited."[37] Despite the outrage this statement caused, in August 2020, the EPA finalized its roll back of Obama's methane rule.[38]

As the expert testimony in *Juliana* establishes, children, including the youth plaintiffs, are bearing serious health burdens caused by fossil fuels. Dr. Jerry Paulson, a renowned expert on the health effects of climate change, explained: "By continuing to promote fossil fuels, the federal government is knowingly putting these children in an increasingly risky situation when it comes to their health."[39] Dr. Paulson finds the defendants' actions "truly shocking" in light of the "undisputed health risks to children."[40] Some plaintiffs are "at risk of irreparable harm from having decreased lung function as a result of growing up in environments with more air pollution."[41] Children who have asthma are already harmed by pollution from fossil fuels, increased prevalence of wildfire smoke, and exacerbated ozone conditions due to climate change. The more fossil fuels are burned, the worse their health will be.[42]

Medical evidence shows children are uniquely vulnerable to psychological harms from climate change.[43] As Dr. Lise Van Susteren, a distinguished psychiatrist with a special interest in the psychological effects of climate change, declared: "Climate change is causing devastating physical impacts—injuries, illnesses, and deaths. But for the magnitude of its impacts, the potential insinuation into every aspect of our lives, the relentlessness of its nature and debilitating effects, it is the emotional toll of climate change that is even more catastrophic, especially for our children. It has the capacity to destroy children psychologically."[44]

In *Juliana*, many of the youth plaintiffs attested to profound impacts to their mental and emotional well-being. Sleeplessness, anxiety, anger, fear, and deep feelings of betrayal by their government are part of the psychological makeup of these young people.[45] As examples of what our youngest generation is experiencing, several youth plaintiffs have described their nightmares, emotional pain, depression, and other psychological harms due to the Trump administration's "institutional betrayal," harms validated by medical experts and worsening each passing day.[46] "Harms that are inflicted intentionally are much more psychologically damaging than what happens to us accidentally. The plaintiffs know that the harm coming to them has been inflicted intentionally and that they are attributable not only to past actions but are also a direct result of actions the federal government is taking today."[47] One youth plaintiff, Aji, put it perfectly: "It increases the pressure cooker feeling that lives in me and ignites my feeling of panic."[48]

GOVERNMENT ACTION

Actions Regarding Renewable Energy Sources

In terms of renewable energy, Trump has long questioned without any evidence the efficacy of renewable energy sources, suggested that wind turbines cause cancer, and championed coal as a clean and efficient source of power. When Trump issued his Executive Order on Promoting Energy Independence and Economic Growth on March 28, 2017,[49] a senior administration official framed the executive order as an "all-of-the-above" energy policy. "We're looking at deposits of coal, looking at nuclear, looking at renewables, all of it."[50] However, the Executive Order really aimed to unleash fossil fuel development by undoing the Obama administration's actions.

The Trump administration has stifled solar and other renewables development on public lands, hindering the country's renewable energy transition.[51] Its budget proposals reflect the decision by the administration to retreat from a federal role in advancing a clean energy economy and maintaining global leadership in the technology.[52] It has proposed repeatedly to slash funding for the Department of Energy's Office of Energy Efficiency and Renewable Energy (EERE), a move that would cripple support for novel and promising technologies for advanced wind turbines, high-tech materials, energy-efficient buildings, and more.[53]

Trump's antipathy toward renewable energy development is reflected in his administration's policies—from imposing tariffs to opposing tax credits—that have cost the sector jobs, stunted growth, and impeded our nation's ability to tackle climate change. The administration repealed the Clean Power Plan (which would have accelerated renewable energy development), imposed tariffs on imported solar cells, and adopted an antagonistic stance on renewable energy tax credits that has stunted the sector's growth.[54] The COVID-19 crisis has amplified the administration's attacks on renewables. The renewable energy industry lost nearly 600,000 jobs—17 percent of all clean energy jobs—in March and April 2020.[55]

In terms of grid stability, there are now at least forty-seven peer-reviewed scientific papers among thirteen independent research groups encompassing ninety-one scientists who find that 100 percent or near 100 percent renewable energy can result in a stable grid at low cost.[56]

Today, even though it is technically and economically feasible to transition swiftly away from fossil energy, and when the climate system is in a dangerous state of emergency, the Trump administration is recklessly increasing fossil fuel development.[57] "The United States is expanding oil and gas extraction on a scale at least four times faster and greater than any other nation and is currently on track to account for 60% of global growth in oil and gas production."[58] As part of the defendants' fossil fuel energy system and strategy for fossil fuel dominance, there are presently close to 100 new fossil fuel infrastructure projects poised for federal permits, including pipelines, export facilities, and coal and liquefied natural gas terminals.[59] Such conduct threatens national security.[60]

TRUMP ADMINISTRATION ENERGY POLICY: OPEN UP FEDERAL LANDS, CLOSE DOWN REGULATIONS

As one of the other experts in *Juliana* wrote: "The federal government has for many years had knowledge, information, and scientific recommendations that it needed to transition the Nation off of fossil fuels in order to first prevent against, and now try to stop, catastrophic climate change. We are well beyond the maxim: 'If you find yourself in a hole, quit digging.'"[61] Despite this admonition, the energy policy of the Trump administration has been to accelerate digging, not quit. Trump began his presidency by introducing his America First Energy Plan, which focuses on expanding fossil fuel production and rolling back investments in wind and solar.[62]

After three years of modest declines, energy-related US CO_2 emissions grew by an estimated 3.4 percent in 2018.[63] This is the second largest annual gain since 1996. The substantial gain in 2018 was due in part to rapid growth in natural gas-fired electricity generation, which saw more than four times the growth of wind and solar generation combined.[64] According to the International Energy Agency (IEA), investment in the United States in new oil and gas wells, pipelines, and other fossil fuel infrastructure, at about $120 billion annually, is greater than in any other country.[65]

Since 2017, the federal government has opened and proposed to open vast additional areas of federal lands (and waters) for leasing for fossil fuel exploration and production. On March 28, 2017, President Trump rescinded the moratorium on coal mining on federal lands.[66] In December 2017, Congress passed and Trump signed legislation authorizing drilling in the Arctic National Wildlife Refuge. In January 2018, as part of President Trump's goal of "Energy Dominance," his administration announced plans to allow new offshore oil and gas drilling in virtually all (98 percent) of US coastal waters during 2019–2024.[67] In February 2018, the DOI's offer of 77 million acres for oil and gas exploration and development (off the coasts of Texas, Louisiana, Mississippi, Alabama, and Florida) was the largest oil and gas sale offering in US history.[68] In November 2018, the DOI announced plans to dramatically expand fossil fuel development in the 23.4-million-acre National Petroleum Reserve–Alaska.[69]

Many of the Trump fossil fuel actions are being challenged in court. Even if some of Trump's efforts to roll back past gains and prevent others fail in the end for one reason or another, precious time and momentum will have been squandered—and indeed squandered at a critical moment in the history of the issue when developments seem to be coming together to offer honest hope.

The Trump White House stands proudly behind its efforts to support the fossil fuel industry, calling this time the "golden era of American energy" supported by "policies that tap into America's incredible energy resources."[70] In particular, in its years in office, the Trump administration has either done or proposed to do the following:

- Repeal the Clean Power Plan (CPP) aimed at reducing emissions from power plants.[71]
- Freeze fuel-efficiency standards for passenger cars and trucks at 37 mpg, rolling back a 2012 rule that would have doubled the fuel economy of passenger cars and trucks to 54 mpg by 2025.[72] On March 30, 2020, Trump finalized the rollback. The EPA's Inspector General has announced it opened an investigation into the agency's weakening of these Obama-era regulations.[73]
- Lift a ban on offshore drilling in many regions, and allow new offshore oil and gas drilling in nearly all United States coastal waters as well as take steps to open the Arctic to oil and gas drilling.[74] On August 17, 2020, the Bureau of Land Management announced its decision to sell drilling rights to the Arctic National Wildlife Refuge's 1.6 million-acre

coastal plain.[75] This auctioning off of vast amounts of onshore and offshore public lands for fossil fuel development is unprecedented.[76]

- Halt the closure of coal plants and bail out failing coal-fired power plants by guaranteeing revenue from the federal government.[77] In June 2020, the DOE announced a $122 million competition for bids to set up "coal innovation centers" throughout the country that aim to develop "value-added, carbon-based products from coal."[78]
- Repeal the methane leak rule.[79] Repealing these controls effectively frees oil and gas companies from the need to detect and repair methane leaks. The Trump rule also makes it harder for the EPA to control methane and other climate pollutants in the future by erecting a new legal hurdle to regulation under a key provision of the Clean Air Act.[80] Also, the Pipeline and Hazardous Material Safety Administration proposed "amendments to the Federal Pipeline Safety Regulations that are intended to ease regulatory burdens on the construction, maintenance and operation of gas transmission, distribution, and gathering pipeline systems."[81]
- Put tariffs on solar panel imports.[82]
- Dissolve the Advisory Committee for the Sustained National Climate Assessment.[83] Instead, in April 2020, Trump formed an "Opening the Country" Energy Council with eight oil and gas executives on this twelve-member energy industry economic task force. There are no representatives from the clean energy sector on the task force.
- Withdraw the United States from the Paris Climate Agreement.[84]
- Reverse course from the past administration and decide that climate change is not a national security threat.[85]
- Withdraw CEQ guidance for federal departments and agencies regarding their consideration of GHG emissions and the impacts of climate change during NEPA (National Environmental Policy Act) reviews.[86] On June 4, 2020, Trump signed an executive order suspending environmental reviews and public comment periods for highways, pipelines, oil and gas projects, and other big infrastructure efforts, and cited the "pressing emergency" caused by the pandemic as a driving force.[87]

- Issue executive orders to revive the Dakota Access and Keystone XL pipelines.[88] On May 22, 2020, the Federal Energy Regulatory Commission reaffirmed its earlier approval of the controversial Jordan Cove liquefied natural gas export terminal, a proposed project that includes construction of a 229-mile pipeline. On July 29, 2020, the DOE's Office of Fossil Energy extended the standard term for authorizations to export natural gas from the lower forty-eight states—including domestically produced liquefied natural gas (LNG), compressed natural gas, and compressed gas liquid—to countries with which the United States does not have a free trade agreement to 2050.[89] This new policy potentially commits future administrations to allowing dozens of firms to export LNG, no matter what climate policies these later administrations seek to enact.
- Numerous other actions to promote fossil fuel development, undermine climate science research, and defund research and development for renewable energy sources, which include revoking an Obama executive order requiring emissions phasedown from federal activities and operations, and repealing regulations reducing emissions of hydrofluorocarbons (HFCs).

CONGRESS: A FAILURE TO LEGISLATE

During his 2016 campaign, Trump announced that his energy policies would address job growth, deregulation, and "energy independence" achieved through the increased production of fossil fuels, including revitalization of the US coal industry. Immediately after the election, Trump, with willing partners in the then Republican-led Congress, worked to fulfill those campaign promises.

Prior to 2018, the Republican Congress took aim at existing regulations by utilizing the Congressional Review Act (CRA).[90] The CRA is an oversight tool that Congress may use to overturn rules issued by federal agencies. The CRA requires agencies to report on their rulemaking activities to Congress and provides Congress with a special set of procedures under which to consider legislation to overturn those rules.[91] These include, among others, the Department of the Interior's Stream Protection Rule (restricting how close coal mines can operate to nearby waterways) and the Securities and Exchange Commission's Resource Extraction Rule requiring publicly traded oil, gas and mining companies to disclose payments to foreign governments.[92] Democrats generally opposed these actions but were powerless to stop them; CRA disapproval resolutions cannot be filibustered in the Senate and thus require only a majority vote.

However, not everything requested by the Trump administration moved through Congress. Congress resisted Trump's efforts to slash the budget of the EPA and fire many of its staff members, and it pushed back on eliminating funding for most research on climate change and renewable energy.[93] The near one-third cuts to the EPA's budget proposed by Trump in early 2017 did not occur; instead, Congress approved cuts of 1 percent only.[94] In 2018, the budget bill also rejected Trump's proposal to eliminate the Advanced Research Projects Agency and instead gave it a 15 percent increase, while maintaining current funding levels for climate change research at NOAA.[95] Trump's December 2017 tax reform bill, which slashed $1.5 trillion from federal taxes, nonetheless retained existing incentives for renewable energy—in part because of support from Republican-led states where clean energy power generation is booming.[96]

Congress reaffirmed the danger of climate change in section 335 of the Defense Authorization Act of 2018.[97] This section states that it is the sense of Congress that "climate change is a direct threat to the national security of the United States and is impacting stability in areas of the world both where the United States Armed Forces are operating today, and where strategic implications for future conflict exist." It also declares that sea-level rise "will threaten the operations of more than 128 United States military sites, and it is possible that many of these at-risk bases could be submerged in the coming years." Finally, section 335 provides, "As global temperatures rise, droughts and famines can lead to more failed states, which are breeding grounds of extremist and terrorist organizations."

Congress is divided and gridlocked, unable to pass major legislation or effectively restrain Trump, who has used executive orders and other powers to fashion energy and environmental policy and regulations. The new majority in the House of Representatives since 2018 has developed a robust set of recommendations for ambitious climate legislation, but the likelihood of it passing both houses of Congress and being signed into law by President Trump is effectively zero.

The most publicized proposal in Congress was the Green New Deal, a resolution drafted by Representative Alexandria Ocasio-Cortez of New York and Senator Ed Markey of Massachusetts.[98] It would establish a select House committee with a mandate to draft aggressive legislation for climate action that would be embedded in a broader US agenda of economic reform, public investment, job creation, and social justice—a Green New Deal that would tackle climate change as the 1930s New Deal tackled the Great Depression.

In January 2019, House Resolution 6 created the bipartisan Select Committee on the Climate Crisis to "develop recommendations on policies, strategies, and innovations to achieve substantial and permanent reductions in pollution and other activities that contribute to the climate crisis."[99] In June 2020, Representative Kathy Castor, Chair of the House Select Committee on the Climate Crisis, and her Select Committee delivered a report containing a framework for congressional action addressed to the scientific and practical imperatives to reduce carbon pollution, make communities more resilient to the impacts of climate change, and build a durable and equitable clean energy economy.[100]

In the House, bipartisan groups of legislators introduced the Climate Solutions Commission Act (H.R. 2326) in 2017 to develop economically viable ways to reduce greenhouse gas emissions and the SUPER Act (H.R. 2858) to reduce pollutants such as black carbon, methane, and hydrofluorocarbons. In the Senate, three Republicans joined with Democrats to protect the Bureau of Land Management's Natural Gas Waste rule, which limits industrial release of harmful methane gas on public and tribal land. Eleven House Republicans also voted to preserve the rule.

Significantly, in 2020, eight senators and fifty-one representatives introduced a Concurrent Resolution in both the US Senate and House of Representatives to protect the fundamental rights of children to a climate system capable of sustaining human life. S. Con. Res. 47 and H. Con. Res.

119, the Children's Fundamental Rights and Climate Recovery, support the young *Juliana* plaintiffs in recognizing that the current climate crisis disproportionately affects the health, economic opportunity, and fundamental rights of children, and demand that the United States develop a national, comprehensive, and science-based recovery plan to meet necessary emissions reduction targets.

As I write, the level of Congressional interest in the climate issue is high, but that is the easy part.

Box 8.1 Keeping Track of Trump

Several organizations have been keeping track of the Trump administration actions that are exacerbating the climate crisis. The resources here provide additional information on actions noted in this expert report.

- Climate Deregulation Tracker—Sabin Center for Climate Change Law at Columbia Law School: https://climate.law.columbia.edu/climate-deregulation-tracker
- The *New York Times* compilation of environmental rollbacks, "The Trump Administration Is Reversing": https://www.nytimes.com/interactive/2020/climate/trump-environment-rollbacks.html
- Regulatory Rollback Tracker—Harvard Environmental & Energy Law Program: http://environment.law.harvard.edu/policy-initiative/regulatory-rollback-tracker
- EPA Mission Tracker—Harvard Environmental & Energy Law Program: https://eelp.law.harvard.edu/epa-mission-tracker/
- *Washington Post* compilation of Trump rollbacks, "How Trump Is Rolling Back Obama's Legacy": https://www.washingtonpost.com/graphics/politics/trump-rolling-back-obama-rules
- Trump Watch: EPA—Environmental Integrity Project: https://environmentalintegrity.org/trump-watch-epa/
- *National Geographic* environmental policy tracker, "A Running List": https://www.nationalgeographic.com/news/2017/03/how-trump-is-changing-science-environment

9 CONCLUSION

Focusing on this history, no matter how extensive its scope, we have to see this period of almost half a century as an instant in time, for that is indeed what it is in Earth's time and in the lives of today's children and all those who must live with its consequences. And in this snapshot of decades, we find a federal government planning for, guiding, supporting, and encouraging massive fossil fuel use despite tragic consequences easily foreseen and avoided.

Actions—commissions and omissions—by the defendants across all presidential administrations have led to US fossil fuel consumption and production increasing significantly since the 1950s (see figures 9.1 and 9.2).

While the United States relied heavily on fossil fuel imports for decades, recently there has been an increase in fossil exports from the United States (see figure 9.3). Actions by the defendants that have perpetuated reliance on fossil fuels have resulted in the release of a massive, and dangerous, amount of CO_2 emissions since 1960 (see figure 9.4). Cumulatively, the United States has emitted more CO_2 than any other nation, and annually, the United States remains the second largest emitter in the world.

After analyzing the last four decades of actions and inactions, a clear pattern of historical government conduct emerges relating to the nation's energy system and climate change. For decades:

a. The defendants have understood both that the dangers of climate change are real, present, and intensifying and that they are caused predominantly by burning fossil fuels.

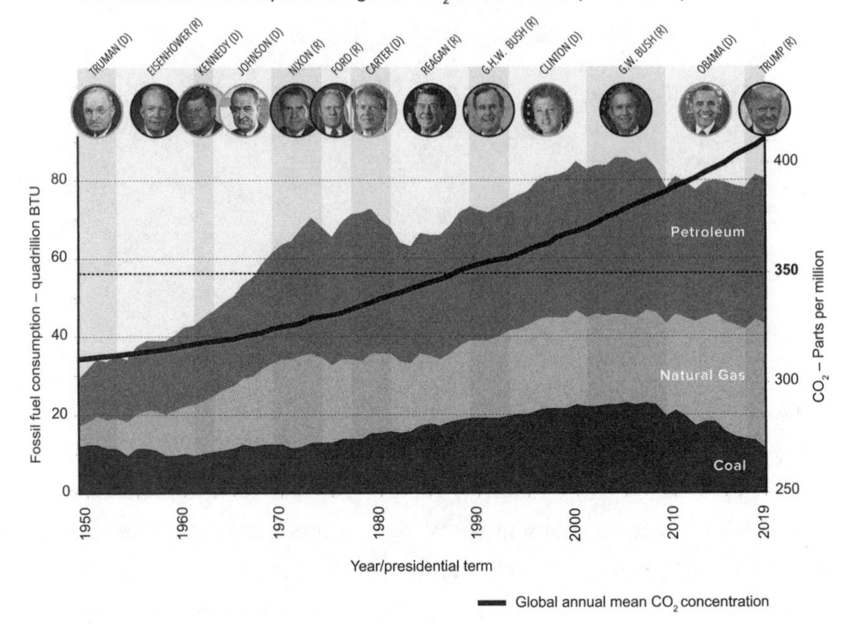

Figure 9.1

US fossil fuel consumption and global CO_2 concentration. Data from US Energy Information Administration, *Monthly Energy Review April 2020*, 2020; Global CO_2 concentration data: 1949–1958: National Aeronautics and Space Administration, Global mean CO_2 mixing ratios, accessed July 2018, https://data.giss.nasa.gov/modelforce/ghgases/Fig1A.ext.txt. 1959–2017: National Oceanic and Atmospheric Administration Earth System Research Laboratory, accessed April 2020, https://www.esrl.noaa.gov/gmd/ccgg/trends/data.html.

b. The defendants have understood how climate change will harm the nation and especially youth plaintiffs and future generations.
c. The defendants have understood there are alternative national energy system pathways that would provide greater protection and safety for the nation and our people.

Notwithstanding these understandings, the defendants have acted routinely and consistently, and continue to do so, to promote fossil fuels and thus to cause irreversible climate danger, a pattern that can only reflect a deliberate indifference to the severe impacts that will follow—impacts to be endured predominantly by youth plaintiffs and future generations.

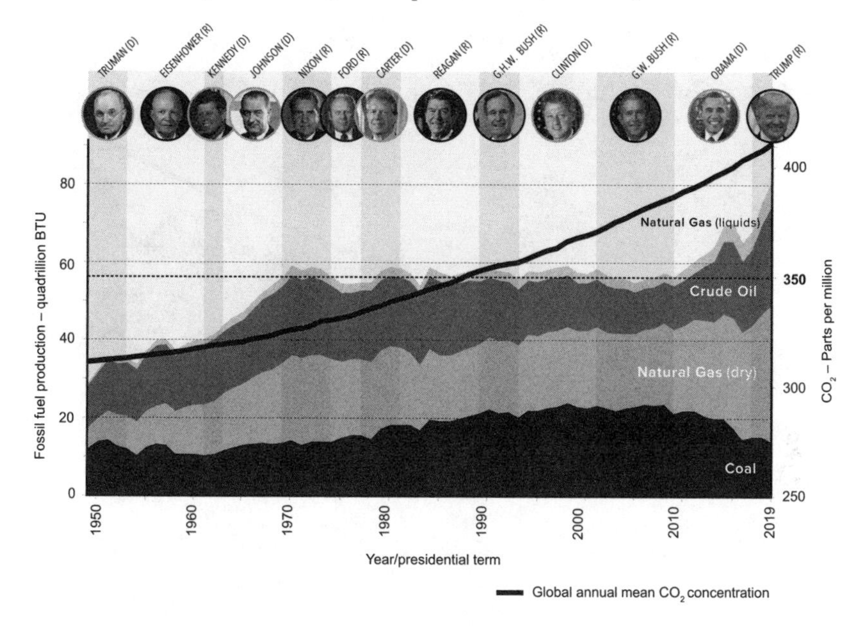

Figure 9.2

US fossil fuel production and global CO_2 concentration. Data from US Energy Information Administration, *Monthly Energy Review April 2020*; Global CO_2 concentration data: 1949–1958: National Aeronautics and Space Administration, Global mean CO_2 mixing ratios; 1959–2017: National Oceanic and Atmospheric Administration Earth System Research Laboratory.

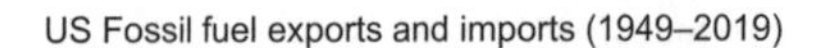

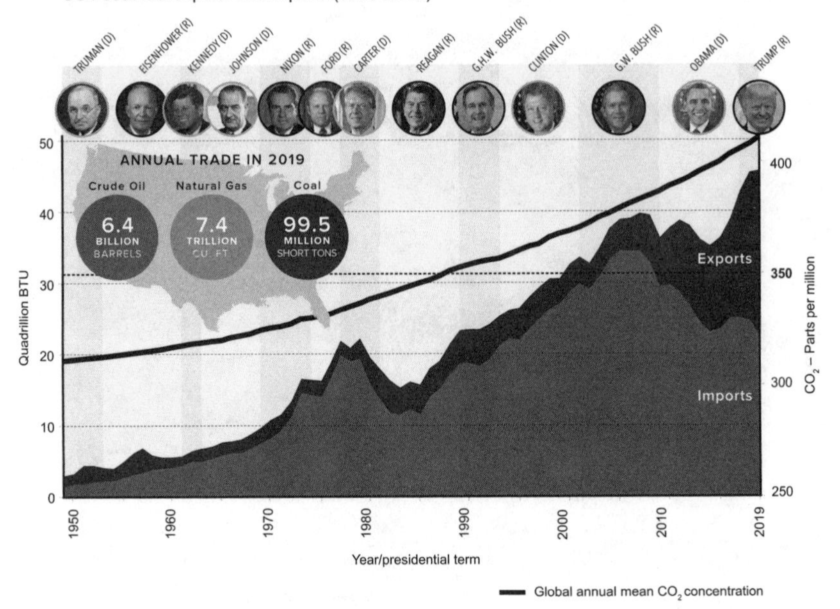

Figure 9.3

US fossil fuel exports and imports. Data from US Energy Information Administration, *Monthly Energy Review April 2020*; Global CO_2 concentration data: 1949–1958: National Aeronautics and Space Administration, Global mean CO_2 mixing ratios; 1959–2017: National Oceanic and Atmospheric Administration Earth System Research Laboratory.

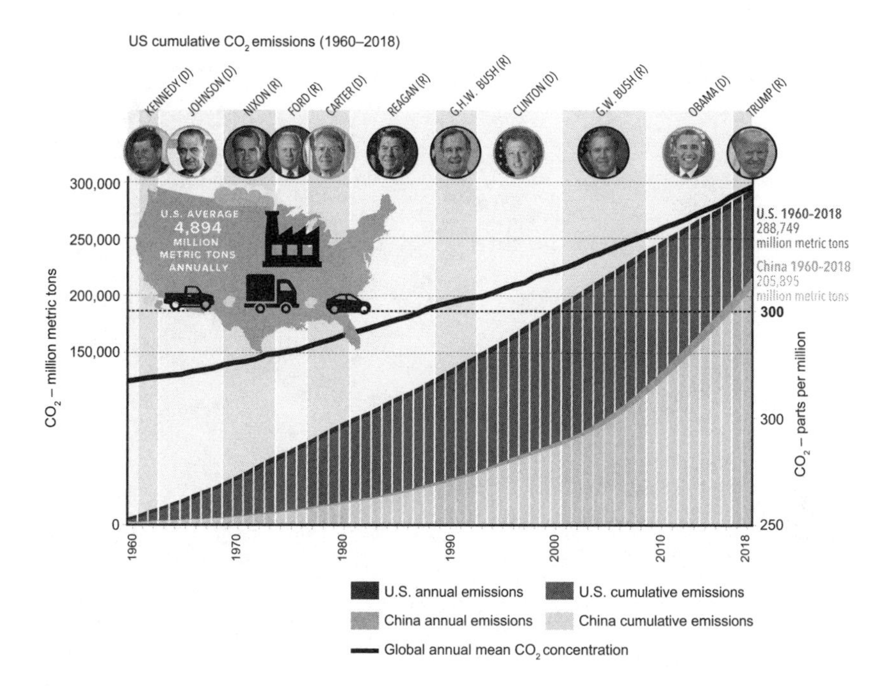

Figure 9.4

US cumulative CO_2 emissions. Data from D. Gilfillan, G. Marland, T. Boden, and R. Andres, *Global, Regional, and National Fossil-Fuel CO_2 Emissions*, Carbon Dioxide Information Analysis Center, Appalachian State University, https://energy.appstate.edu/research/work-areas/cdiac-appstate (accessed through globalcarbonatlas.org, August 2020); Global CO_2 concentration data: National Oceanic and Atmospheric Administration Earth System Research Laboratory.

Appendix: The Procedural History of *Juliana*

Julia Olson and Philip Gregory

The twenty-one courageous youth plaintiffs, along with a youth organization known as Earth Guardians, and Dr. James Hansen as guardian for Future Generations, commenced *Juliana* on August 12, 2015, by filing a Complaint in the US District Court of Oregon in Eugene. The youths' Complaint asserts that through the federal government's affirmative actions that cause climate change, it has violated the youngest generation's constitutional rights to life, liberty, property, and equal protection of the law, as well as failed to protect essential public trust resources. The Complaint alleges that the defendants' systemic affirmative ongoing conduct, persisting over decades, in creating, controlling, and perpetuating a national fossil fuel–based energy system, despite long-standing knowledge of the resulting destruction to our nation and profound harm to these young plaintiffs, violates the youths' constitutional due process and equal protection rights. Specifically, the plaintiffs allege:

- Defendants' conduct violates their substantive due process rights to life, liberty, and property, to dignity, to personal security, to a stable climate system capable of sustaining human lives and liberties, as well as other previously recognized unenumerated liberty interests, and has placed plaintiffs in a position of danger with deliberate indifference to their safety under a state-created danger theory.
- Defendants' conduct violates their rights as children to equal protection by discriminating against them with respect to their fundamental rights and as members of a protected or quasi-protected class.

- Defendants' conduct violates their rights as beneficiaries to public trust resources under federal control and management.

The Complaint seeks a declaration of the plaintiffs' rights and the violation thereof and an order directing the defendants to cease their violations of the plaintiffs' rights; to prepare an accounting of the nation's GHG emissions; and to prepare and implement an enforceable national remedial plan to cease the constitutional violation by phasing out fossil fuel emissions and drawing down excess atmospheric CO_2, as well as such other and further relief as may be just and proper.

While we only sued agencies and departments of the federal government and their heads, the trade associations for the fossil fuel industry initially intervened in the case as defendants, joining the US government in filing motions to try to have the case dismissed—a true David versus Goliath battle. In April 2016, US Magistrate Judge Thomas Coffin recommended denial of both motions to dismiss. On November 10, 2016, US District Court Judge Ann Aiken upheld Judge Coffin's recommendation, issuing a historic opinion and order denying the motions. Among her rulings, Judge Aiken found:

- "This action is of a different order than the typical environmental case. It alleges that [the federal government's] actions and inactions—whether or not they violate any specific statutory duty—have so profoundly damaged our home planet that they threaten plaintiffs' fundamental constitutional rights to life and liberty."
- "Plaintiffs have alleged that defendants played a significant role in creating the current climate crisis, that defendants acted with full knowledge of the consequences of their actions, and that defendants have failed to correct or mitigate the harms they helped create in deliberate indifference to the injuries caused by climate change. They may therefore proceed with their substantive due process challenge to defendants' failure to adequately regulate CO_2 emissions."
- "I have no doubt that the right to a climate system capable of sustaining human life is fundamental to a free and ordered society."

- "Where a complaint alleges government action is affirmatively and substantially damaging the climate system in a way that will cause human deaths, shorten human lifespans, result in widespread damage to property, threaten human food sources, and dramatically alter the planet's ecosystem, it states a claim for a due process violation. To hold otherwise would be to say that the Constitution affords no protection against a government's knowing decision to poison the air its citizens breathe or the water its citizens drink."

On January 13, 2017, the defendants filed their Answer, admitting many of plaintiffs' scientific and factual allegations. In their Answer, the defendants admit they affirmatively "permit, authorize, and subsidize fossil fuel extraction, development, consumption, and exportation"; that emissions "from such activities have increased the atmospheric concentration of CO_2"; that the United States is responsible for more than a quarter of global historic cumulative CO_2 emissions; and that "current and projected atmospheric concentrations of six well-mixed GHGs, including CO_2, threaten the public health and welfare of current and future generations, and this threat will mount over time as GHGs continue to accumulate in the atmosphere and result in ever greater rates of climate change."

In June 2017, by filing a rare "petition for writ of mandamus" with the Ninth Circuit Court of Appeals, the Trump administration began to employ its drastic tactics to keep *Juliana* from going to trial, to silence the voices of youth, and to keep science out of the courtroom. In July, the Ninth Circuit requested that we submit a response to the government's petition for writ of mandamus. At the time we filed our response, eight *amicus* ("friend of the court") briefs were filed with the Ninth Circuit in support of *Juliana*. On December 11, 2017, a three-judge panel of the Ninth Circuit, made up of Chief Judge Sidney Thomas and Circuit Judges Alex Kozinski and Marsha Berzon, heard oral arguments. Julia Olson argued to a courtroom packed with those in favor of the youth plaintiffs; Eric Grant argued on behalf of the Trump administration. Judge Kozinski subsequently resigned from the Ninth Circuit and was replaced on the panel by Circuit Judge Michelle Friedland.

On March 7, 2018, the Ninth Circuit rejected the Trump administration's "drastic and extraordinary" petition for writ of mandamus.[1]

During the April 12, 2018 case management conference, October 29, 2018, was set as the start date for our trial. On July 18, 2018, Judge Aiken heard oral arguments in our case as part of the Trump administration's latest procedural tactics to avoid trial: a motion for judgment on the pleadings and a motion for summary judgment. In opposition to these motions, we submitted eighteen expert declarations, twenty-one plaintiff declarations, and hundreds of government documents into the record, totaling more than 36,000 pages. For oral argument, our supporters packed the Wayne Morse Federal Courthouse courtroom in Eugene and three overflow courtrooms.

We want to briefly discuss one of the key parts of the federal government's defense heading into trial. Remarkably, the defendants contended that the federal government can knowingly deprive US children of a life-sustaining climate system, *the very foundation of all life*, without violating the Constitution. By positing that the climate right is not fundamental, the defendants asserted they have the ongoing power to deliberately alienate, infringe upon, and deprive these children of a livable future without meaningful due process. As you will see by reading this book, without the judiciary's check on the executive branch and its assertion of unrestrained power, that is precisely what the defendants will continue doing to these children and future generations.

The defendants' groundless arguments would dismantle the concept of unalienable rights settled since the Declaration of Independence. Unalienable rights, such as "Life, Liberty and the pursuit of Happiness," are natural rights, not bestowed by the laws of people, but "endowed by their Creator."[2] The right of these youth plaintiffs to live with the climate system that nature provides, "endowed by their Creator," free of government-sanctioned destruction, is the very foundation of, and preservative of, all of their unalienable natural rights. It is, in fact, the prerequisite to life itself. As James Madison said in 1818, "Animals, including man, and plants may be regarded as the most important part of the terrestrial creation. . . . To all of them, the atmosphere is the breath of life. Deprived of it, they all equally perish."[3]

Before Judge Aiken could decide these motions, and continuing its efforts to keep the case from going to trial, the Trump administration filed its second petition for writ of mandamus with the Ninth Circuit. On July 20, 2018, the court again rejected the Trump administration's petition.

Not to be deterred, the Trump administration applied for a stay with the United States Supreme Court. On July 30, the Supreme Court ruled in our favor, one of Justice Anthony Kennedy's last acts from the bench before retirement. The Supreme Court denied the Trump administration's application for stay, preserving the start date for the trial of October 29, 2018. The Supreme Court also denied the government's "premature" request to review the case before the district court had heard all of the facts that support the youths' claims at trial.

Throughout the summer of 2018, not only were we beating back the Trump administration's incessant efforts to stop the October 29 trial, but our team was engaged in taking and defending numerous depositions. Each of the twenty-one expert witnesses for the plaintiffs was deposed. The Trump administration also chose to notice the depositions of each of the youth plaintiffs.

On October 15, Judge Aiken granted the Trump administration's motions for judgment on the pleadings and summary judgment in part, by limiting the scope of the plaintiffs' claims and dismissing the president from the case without prejudice, thereby narrowing the plaintiffs' case.[4]

Regarding the plaintiffs' due process claim, the district court found that the "plaintiffs have introduced sufficient evidence and experts' opinions to demonstrate a question of material fact as to federal defendants' knowledge, actions, and alleged deliberate indifference."[5]

On October 18, the Trump administration filed a second writ of mandamus petition and application for stay with the US Supreme Court, asking it to circumvent the ordinary procedures of federal litigation and stop the trial of *Juliana*. On October 19, Chief Justice Roberts ordered a temporary stay of discovery and trial while the full court considered the federal government's petition and asked us to file a response.[6] On October 22, we responded, requesting that the Court allow trial to proceed on October 29. On November 2, the Supreme Court denied the government's second application and lifted the temporary stay.[7]

However, we had lost our October 29 trial date. Even though there was no trial, on October 29, thousands rallied at courthouses around the country to support our clients' right to be heard at trial.

On November 21, Judge Aiken issued an order certifying four orders in our case for interlocutory appeal to the Ninth Circuit and staying all proceedings under the Interlocutory Appeals Act, 28 U.S.C. §1292(b). In doing so, Judge Aiken set forth the many reasons why she believed interlocutory appeal was not appropriate. Judge Aiken also indicated the court "stands by its prior rulings on jurisdictional and merits issues, as well as its belief that this case would be better served by further factual development at trial."

On November 30, the defendants petitioned the Ninth Circuit for permission to appeal the certified orders. On December 10, we answered, opposing certification. We outlined the further delay that would occur if there was an interlocutory appeal, the urgent nature of the case and the climate crisis, and the likely need for preliminary injunctive relief should interlocutory appeal be awarded to the defendants. On December 26, a panel of the Ninth Circuit granted the defendants' petition for permission to bring an interlocutory appeal[8] over a sharp dissent from Circuit Judge Friedland, which stated in part:

- "Although the district court's statement that the [28 U.S.C.] §1292(b) factors were met would ordinarily support certification, here it appears that the court felt compelled to make that declaration even though—as the rest of its order suggests—the court did not believe that to be true. This is very concerning."
- "It is also concerning that allowing this appeal now effectively rewards the Government for its repeated efforts to bypass normal litigation procedures by seeking mandamus relief in our court and the Supreme Court. If anything has wasted judicial resources in this case, it was those efforts."[9]

On February 7, 2019, we moved for a preliminary injunction, seeking to prevent the federal government from issuing leases and mining permits for extracting coal on federal public lands, leases for offshore oil and gas exploration and extraction activities, and federal approvals for new fossil fuel infrastructure. Because our case was then before the Ninth Circuit, we took the unusual step of filing that motion in the Court of Appeals while the government's interlocutory appeal was being heard.

On February 19, Zero Hour, a youth-led climate group launched the website www.joinjuliana.org and announced its nationwide campaign to help tens of thousands of young people add their names to the Young People's *amicus* brief in support of our youth plaintiffs. On March 1, more powerful voices of support for our case filed *amicus curiae* briefs with the Ninth Circuit. In all, fifteen *amicus* briefs, filed on behalf of a diverse set of supportive communities, including members of the US Congress, legal scholars, religious and women's groups, businesses, historians, medical doctors, international lawyers, environmentalists, and more than 32,000 youth under the age of twenty-five, displayed legal support for *Juliana* to proceed to trial.[10]

On March 24, the Ninth Circuit set June 4, 2019, to hear oral arguments in Portland, Oregon, on the interlocutory appeal and the preliminary injunction. At the hearing on June 4, Julia Olson argued on our behalf and Assistant Attorney General Jeffrey Clark argued on behalf of the federal government before US Circuit Judges Mary H. Murguia and Andrew D. Hurwitz of the Ninth Circuit, and District Judge Josephine L. Staton of the District Court for the Central District of California, sitting by designation.

On January 17, 2020, a divided panel of the Ninth Circuit issued its order on the interlocutory appeal.[11] In its ruling, the majority recognized the gravity of the evidence on the plaintiffs' injuries from climate change and the government's role in causing them. Yet while it found the government is violating the plaintiffs' constitutional rights, two of the three judges said the executive and legislative branches should address the youth plaintiffs' requested remedies, not the judiciary. In ordering the case dismissed, the majority opinion found:

- "A substantial evidentiary record documents that the federal government has long promoted fossil fuel use despite knowing that it can cause catastrophic climate change, and that failure to change existing policy may hasten an environmental apocalypse."
- "Atmospheric carbon dioxide has skyrocketed to levels not seen for almost three million years."
- "The country is now expanding oil and gas extraction four times faster than any other nation."
- "The problem is approaching 'the point of no return.' Absent some action, the destabilizing climate will bury cities, spawn life-threatening natural disasters, and jeopardize critical food and water supplies."
- The majority agreed with Judge Staton's dissent on "the gravity of the plaintiffs' evidence; we differ only as to whether an Article III court can provide their requested redress."
- "It is beyond the power of an Article III court to order . . . the plaintiffs' requested remedial plan."
- "The [children's] impressive case for redress must be presented to the political branches of government" . . . "or to the electorate at large . . . through the ballot box."

In her dissent, Judge Staton stated she would have affirmed the plaintiffs' constitutional climate rights, writing that "our nation is crumbling—at our government's own hand—into a wasteland." Judge Staton went on to assert:

- "Seeking to quash this suit, the government bluntly insists that it has the absolute and unreviewable power to destroy the Nation. My colleagues throw up their hands, concluding that this case presents nothing fit for the Judiciary."
- Fundamental rights "may not be submitted to vote; they depend on the outcome of no elections."
- They are "beyond the reach of majorities and officials."
- "Such relief, much like the desegregation orders and statewide prison injunctions the Supreme Court has sanctioned, would vindicate plaintiffs' constitutional rights without exceeding the Judiciary's province."

- "When the seas envelop our coastal cities, fires and droughts haunt our interiors, and storms ravage everything between, those remaining will ask: Why did so many do so little?"

On March 2, 2020, we filed a petition for rehearing *en banc* with the Ninth Circuit. This petition asks the full Ninth Circuit Court of Appeals to convene a new panel of eleven circuit court judges to review the January 17 sharply divided opinion. The federal government opposed our petition.

On March 12, twenty-four members of the US Congress, prominent experts in the fields of constitutional law, climate change, and public health, and several leading women's, children's, environmental, and human rights organizations filed ten *amicus curiae* briefs with the Ninth Circuit in support of *Juliana*.[12] The *amicus* briefs urge the Ninth Circuit to grant the plaintiffs' *en banc* petition.

As of the writing of this appendix, we are awaiting a decision from the Ninth Circuit on the *en banc* petition and the entrance of a third set of federal defendants with the incoming Biden–Harris administration. Will the new administration come to the settlement table and support the youths' standing in *Juliana*, or will it follow the path of its predecessors? The youth will soon learn whether the new administration will stand up for their constitutional rights and their access to our Article III courts and stop perpetuating the climate emergency.

Qualifications of the Author

I, James Gustave Speth, am a retired professor of law at the Vermont Law School and a senior fellow at Vermont Law School, the Democracy Collaborative, and the Tellus Institute. During the Carter administration I served as member and chairman of the US Council on Environmental Quality. After leaving government, I founded and for a decade was president of the World Resources Institute. I also taught environmental and constitutional law at Georgetown University Law Center. From 1993 to 1999, I was Administrator of the United Nations Development Programme and served as chair of the UN Development Group. I served as Dean of the Yale School of Forestry and Environmental Studies from 1999 to 2009. From 1970 to 1977 I was a senior attorney with the Natural Resources Defense Council, which I helped to found.

I have a long career of working at the intersection of government and environmental protection, including providing leadership to many task forces and committees whose roles have been to combat environmental degradation, including the President's Task Force on Global Resources and Environment, the Western Hemisphere Dialogue on Environment and Development, and the National Commission on the Environment. Among my awards are the National Wildlife Federation's Resources Defense Award, the Natural Resources Council of America's Barbara Swain Award of Honor, a 1997 Special Recognition Award from the Society for International Development, lifetime achievement awards from the Environmental Law Institute and the League of Conservation Voters, and the

Blue Planet Prize. I hold honorary degrees from Clark University, the College of the Atlantic, the Vermont Law School, Middlebury College, the University of South Carolina, Unity College, and the University of Massachusetts Boston.

In preparing my expert report and testifying at trial, I am not receiving any compensation and am providing my expertise *pro bono* to the plaintiffs.

Abbreviations

ANWR	Arctic National Wildlife Refuge
AR5	Fifth Assessment Report
BBL/D	barrels per day
BLM	Bureau of Land Management
CAA	Clean Air Act
CAFE	corporate average fuel economy
CBO	Congressional Budget Office
CCRI	Climate Change Research Initiative
CCSP	Climate Change Science Program
CEQ	Council on Environmental Quality
CO_2	carbon dioxide
COP	Conference of the Parties
CPP	Clean Power Plan
CRS	Congressional Research Service
DARPA	Defense Advanced Research Projects Agency
DPC	Domestic Policy Council
DOE	Department of Energy
DOI	Department of the Interior
DOT	Department of Transportation
EIA	Energy Information Administration

EO	Executive Order
EPA	Environmental Protection Agency
EPACT	Energy Policy Act
FERC	Federal Energy Regulatory Agency
GAO	General Accounting Office (changed to Government Accountability Office in 2004)
GARP	Global Atmospheric Research Program
GCRP	Global Change Research Program
GDP	gross domestic product
GFDL	Geophysical Fluid Dynamic Laboratory
GHG	greenhouse gas or gases
IPCC	Intergovernmental Panel on Climate Change
NAS	National Academy of Sciences
NASA	National Aeronautics and Space Administration
NCAR	National Center for Atmospheric Research
NCPO	National Climate Program Office
NEPA	National Environmental Policy Act
NGL	natural gas liquids
NOAA	National Oceanic and Atmospheric Administration
NRC	National Research Council
NSF	National Science Foundation
OTA	Congressional Office of Technological Assessment
OCS	Outer Continental Shelf
OST	Office of Science and Technology
OSTP	Office of Science and Technology Policy
PPM	parts per million
PSAC	President's Science Advisory Committee
Q&A	question and answer

R&D	research and development
SERI	Solar Energy Research Institute
UN	United Nations
UNFCC	United Nations Framework Convention on Climate Change
USGS	US Geological Survey
WMO	World Meteorological Organization
WRI	World Resources Institute

Reader Access to Exhibits

In 2018, Gus Speth originally prepared much of the text for this book as his Expert Report for the children's constitutional climate lawsuit, *Juliana v. United States*. As is often the case in federal court litigation, his Expert Report referred to exhibits that were to be used at trial in the *Juliana* litigation. His book achieves its impact, in part, by a detailed presentation of original source material drawing on extensive research of the record over the past fifty years, mainly, but not entirely, government documents stretching from Presidents Johnson through Trump. These memos, reports, and documents provide an opportunity for readers to examine some of the federal government's knowledge and actions in causing the current climate crisis.

This book attempts to make its sources as transparent as possible. Selected exhibits from the Speth Expert Report for *Juliana v. United States* are digitally available through this book's page at the MIT Press website: https://mitpress.mit.edu/books/they-knew. The exhibits also may be accessed through the Our Children's Trust website: https://www.ourchildrenstrust.org/.

Notes

INTRODUCTION

1. Climate change has occurred naturally throughout the ages, and continues to occur without human interference. However, over the past century, events caused by human activities or forcings have become the most significant contribution to climate change. Today, climate change occurs to a large and measurable extent through human production of warming gases, called greenhouse gases, and has escalated the heating of the planet to the point of a global emergency.
2. OCT does not advocate for a right to particular climate conditions. The climate system naturally varies across time. Nor does OCT assert a right to live in a "pollution-free" or "healthy environment." OCT seeks to protect the climate system from government-sanctioned impairment resulting in dangers to the lives, liberties, and property of children and future generations.
3. The climate system is made up of Earth's atmosphere (our gas-composed air), hydrosphere (our freshwater and oceans), cryosphere (our ice), biosphere (our living ecology and organisms like trees), and lithosphere (our land and soils). NOAA, Geophysical Fluid Dynamics Laboratory, Develop Improved and More Comprehensive Earth System Models, https://www.gfdl.noaa.gov/climate-and-ecosystems-comprehensive-earth-system-models; NOAA, "National Weather Service Glossary 'C,'" https://www.weather.gov/ggw/GlossaryC.
4. Declaration of Dr. James E. Hansen in Support of Plaintiffs' Urgent Motion under Circuit Rule 27–3(b) for Preliminary Injunction, para. 6 (hereinafter Hansen Decl.).
5. Declaration of Steven W. Running in Support of Plaintiffs' Urgent Motion under Circuit Rule 27–3(b) for Preliminary Injunction, para. 4 (hereinafter Running Decl.).
6. See also Declaration of Eric Rignot, Ph.D. in Support of Plaintiffs' Urgent Motion under Circuit Rule 27–3(b) for Preliminary Injunction, para. 15 (hereinafter Rignot Decl.); Hansen Decl., paras. 77–81.
7. See also Running Decl., para. 46; Declaration of Peter A. Erickson in Support of Plaintiffs' Urgent Motion under Circuit Rule 27–3(b) for Preliminary Injunction, paras. 10–11 (hereinafter Erickson Decl.); Hansen Decl., paras. 82–83.

8. Declaration of Joseph E. Stiglitz, Ph.D., in Support of Plaintiffs' Urgent Motion under Circuit Rule 27–3(b) for Preliminary Injunction, para. 8 (hereinafter Stiglitz Decl.).

9. Stiglitz Decl., para. 9.

10. *Brown v. Plata*, 563 U.S. 493 (2011).

11. *Urgenda Foundation v The State of The Netherlands (Ministry of Infrastructure and the Environment)* [2015] Case C/09/456689/HA_ZA 13–1396 (District Court of The Hague, Chamber for Commercial Affairs) http://blogs2.law.columbia.edu/climate-change-litigation/wp-content/uploads/sites/16/non-us-casedocuments/2015/20150624_2015-HAZA-C0900456689_decision-1.pdf, accessed 13 November 2019 (Urgenda Foundation v The State of The Netherlands—Lower Court Decision). In October 2018, the Court of Appeal for The Hague affirmed the District Court decision: see [2018] Case 200.178.245/01, www.urgenda.nl/wp-content/uploads/ECLI_NL_GHDHA_2018_2610.pdf, for the District Court judgment. The Netherlands Supreme Court decision of December 20, 2019 dismissed a further government appeal, and essentially affirmed the Court of Appeal's decision while providing further elaboration as to why the arguments and defenses of the Government were not availing. The summary by the Supreme Court of its decision and the full text of the decision, in English, is found at https://uitspraken.rechtspraak.nl/inziendocument?id=ECLI:NL:HR:2019:2007.

12. Stiglitz Decl., para. 10 ("For decades, the U.S. government has had extensive knowledge that there were viable alternatives to a fossil fuel-based, national energy system, and with the appropriate allocation of further resources to research and development, it is likely that these alternatives would have been even more competitive than fossil fuels"); Declaration of James H. Williams in Support of Plaintiffs' Urgent Motion under Circuit Rule 27–3(b) for Preliminary Injunction, paras. 13–23 (hereinafter Williams Decl.); Declaration of Mark Z. Jacobson in Support of Plaintiffs' Urgent Motion under Circuit Rule 27–3(b) for Preliminary Injunction, paras. 7–19 (hereinafter Jacobson Decl.); Erickson Decl., para. 14 (Defendants have plans for "new offshore oil and gas drilling in virtually all (98%) of U.S. coastal waters during 2019–2024").

13. Erickson Decl., para. 15.

14. *Id.* para. 18.

15. Declaration of Vice Admiral Lee Gunn, USM (Ret.), in Support of Plaintiffs' Urgent Motion under Circuit Rule 27–3(b) for Preliminary Injunction, passim.

16. Stiglitz Decl. para. 9.

17. *Id.* para. 9 n.4.

18. Declaration of Jerome A. Paulson in Support of Plaintiffs' Urgent Motion under Circuit Rule 27–3(b) for Preliminary Injunction, paras. 32–41 (hereinafter Paulson Decl.).

19. Declaration of Journey Z. in Support of Plaintiffs' Urgent Motion under Circuit Rule 27–3(b) for Preliminary Injunction, paras. 10–13 (hereinafter Journey Decl.).

20. Declaration of Levi D. in Support of Plaintiffs' Urgent Motion under Circuit Rule 27–3(b) for Preliminary Injunction, paras. 18–22 (hereinafter Levi Decl.); Journey Decl., paras. 14–19; Paulson Decl., paras. 19–20.

21. Levi Decl., paras. 7, 18, 22; Declaration of Kevin E. Trenberth in Support of Plaintiffs' Urgent Motion under Circuit Rule 27–3(b) for Preliminary Injunction, para. 12 (hereinafter Trenberth Decl.).

22. Paulson Decl. paras. 23, 34, 41.

23. Declaration of Nicholas V. in Support of Plaintiffs' Urgent Motion under Circuit Rule 27–3(b) for Preliminary Injunction, paras. 4–7 (hereinafter Nicholas Decl.); Paulson Decl., paras. 27–30.

24. Running Decl., para. 36.

25. Paulson Decl. para. 23.

26. Declaration of Lise Van Susteren in Support of Plaintiffs' Urgent Motion under Circuit Rule 27–3(b) for Preliminary Injunction, para. 12 (hereinafter Van Susteren Decl.).

27. *Id.*

28. Declaration of Aji P. in Support of Plaintiffs' Urgent Motion under Circuit Rule 27–3(b) for Preliminary Injunction, paras. 3–4, 8, 11; Levi Decl., paras. 7–9, 25; Journey Decl., paras. 25–26; Nicholas Decl., para. 8.

29. Van Susteren Decl., paras. 21–28; Paulson Decl., paras. 39–42; Declaration of Julia A. Olson in Support of Plaintiffs' Urgent Motion under Circuit Rule 27–3(b) for Preliminary Injunction, Ex. 2, 10–11 (Frumkin Report) (hereinafter Olson Decl.).

30. Van Susteren Decl., para. 20.

31. *Collins v. City of Harker Heights*, 503 U.S. 115, 125 (1992).

32. *Washington v. Glucksberg*, 521 U.S. 702, 721 (1997) (quoting *Reno v. Flores*, 507 U.S. 292, 302 (1993)).

33. *Youngberg v. Romeo*, 457 U.S. 307, 315–16 (1982).

34. See *Youngberg*, 457 U.S. at 315 ("In the past, this Court has noted that the right to personal security constitutes a 'historic liberty interest' protected substantively by the Due Process Clause" (quoting *Ingraham v. Wright*, 430 U.S. 651, 673 (1977))).

35. *DeShaney v. Winnebago Cnty. Dep't of Soc. Servs.*, 489 U.S. 189, 197 (1989).

36. *Id.* at 197.

37. *Id.* at 198.

38. *Id.* at 199–200.

39. *Id.* at 201.

40. See *Butera v. District of Columbia*, 235 F.3d 637, 648–49, 649 n.10 (D.C. Cir. 2001) (collecting cases).

41. *Pauluk v. Savage*, 836 F.3d 1117, 1122 (9th Cir. 2016) (citations omitted).

42. *L.W. v. Grubbs (Grubbs II)*, 92 F.3d 894, 900 (9th Cir. 1996).

43. *Patel v. Kent Sch. Dist.*, 648 F.3d 965, 974 (9th Cir. 2011) (quoting *Bryan Cty. v. Brown*, 520 U.S. 397, 410 (1997)).

44. *Pauluk*, 836 F.3d at 1124 (quoting *Patel*, 648 F.3d at 974).

45. *Hernandez v. City of San Jose*, 897 F.3d 1125, 1134–35 (9th Cir. 2018); *DeShaney v. Winnebago Cnty. Dep't of Soc. Servs.*, 489 U.S. 189, 196 (1989).

46. *Id.* (quoting *Grubbs II*, 92 F.3d at 899).

47. *Id.* (quoting *Grubbs II*, 92 F.3d at 900).

48. Hansen Decl., para. 51

49. Appellants' Opening Brief, 40 (filed February 1, 2019).

50. Hansen Decl., para. 38–55, Ex. 1, 26, 41–43; Declaration of Ove Hoegh-Guldberg in Support of Plaintiffs' Urgent Motion under Circuit Rule 27–3(b) for Preliminary Injunction, paras. 16–23; Erickson Decl., para. 28, *passim*; Olson Decl., Ex. 5 (Wanless Report).

51. Hansen Decl., para. 35; Erickson Decl., para. 8 ("energy-related U.S. CO_2 emissions from fossil fuel combustion grew by about 3.4% in 2018").

52. Hansen Decl., paras. 35–37; Erickson Decl., para. 28, *passim.*

53. D. Ct. Doc. 98, paras. 7, 213.

54. Erickson Decl., para. 14.

55. *DeShaney*, 489 U.S. at 196; Hansen Decl., para. 9

56. Hansen Decl., para. 9.

57. *County of Sacramento v. Lewis*, 523 U.S. 833, 853 (1998) ("When such extended opportunities to do better are teamed with protracted failure even to care, indifference is truly shocking"); Williams Decl., para. 23; Jacobson Decl., paras. 7–13 (describing how transitioning to 100% renewable energy will cost less than the current fossil fuel–based energy system).

58. See *Farmer v. Brennan*, 511 U.S. 825, 836 (1994) ("acting or failing to act with deliberate indifference to a substantial risk of serious harm . . . is the equivalent of recklessly disregarding that risk"); Trenberth Decl., para. 13 (calling Defendants' actions "extremely reckless"); Erickson Decl., para. 15 ("It is my opinion that expanding U.S. fossil fuel extraction is a reckless course of conduct"); Hansen Decl., para. 82; Stiglitz Decl., para. 9; Ex. 1, paras. 40.

59. *Hernandez*, 897 F.3d at 1136.

60. *Patel*, 648 F.3d at 971–972 (citation and internal quotation marks omitted).

61. Pauli Murray, *Pauli Murray: The Autobiography of a Black Activist, Feminist, Lawyer, Priest, and Poet* (University of Tennessee Press, 1989).

62. *Plessy v. Ferguson*, 163 U.S. 537 (1896).

63. *Brown v. Board of Ed. of Topeka, Shawnee Cnty, Ka.*, 347 U.S. 483 (1954).

64. *Westminster Sch. Dist. of Orange Cnty v. Mendez*, 161 F.2d 774 (1947).

65. Jack Bass, *Unlikely Heroes* (University of Alabama Press, 1990).

66. Pauli Murray, *Dark Testament and Other Poems*, v.8 (Liveright, 2018).

67. Gus Speth, *What We Have Instead* (Manchester Center, VT: Shires Press, 2019); Gus Speth, *It's Already Tomorrow* (Manchester Center, VT: Shires Press, 2020).

68. Conclusion, this volume.

69. Conclusion, this volume.

70. Those individuals serving *pro bono* as expert witness on behalf of Plaintiffs include Frank Ackerman, Peter Erickson, Howard Frumkin, James Hansen, Ove Hoegh-Guldberg, Mark Jacobson, Akilah Jefferson, Susan Pacheco, Jerome Paulson, Eric Rignot, G. Philip Robertson, Steve Running, Catherine Smith, Gus Speth, Joseph Stiglitz, Kevin Trenberth, Lise Van Susteren, Karrie P. Walters, Harold Wanless, James H. Williams, and Andrea Wulf.

OVERVIEW

1. The following definitions apply as of September 2020 for purposes of this report:

The term "plaintiffs" means: KELSEY CASCADIA ROSE JULIANA; XIUHTEZCATL TONATIUH MARTINEZ; ALEXANDER LOZNAK; JACOB LEBEL; ZEALAND B., through his Guardian Kimberly Pash-Bell; AVERY M., through her Guardian Holly McRae; SAHARA V., through her Guardian Toa Aguilar; KIRAN ISAAC OOMMEN; TIA MARIE HATTON; ISAAC VERGUN; MIKO VERGUN; HAZEL V., through her Guardian Margo Van Ummersen; SOPHIE KIVLEHAN; JAIME BUTLER; JOURNEY ZEPHIER; VIC BARRETT; NATHANIEL BARING; AJI PIPER; LEVI D., through his Guardian Leigh-Ann Draheim; JAYDEN F., through her Guardian Cherri Foytlin; NICHOLAS VENNER; EARTH GUARDIANS, a nonprofit organization; FUTURE GENERATIONS, through their Guardian Dr. James Hansen.

The term "defendants" means: UNITED STATES OF AMERICA; DONALD J. TRUMP, in his official capacity as President of the United States; OFFICE OF THE PRESIDENT OF THE UNITED STATES; MARY B. NEUMAYR, in her capacity as Chairman of Council on Environmental Quality; RUSSELL VOUGHT, in his official capacity as Director of the Office of Management and the Budget; KELVIN K. DROEGEMEIR, in his official capacity as Director of the Office of Science and Technology Policy; U.S. DEPARTMENT OF ENERGY; DAN BROUILLETTE, in his official capacity as Secretary of Energy; U.S. DEPARTMENT OF THE INTERIOR; DAVID L. BERNHARDT, in his official capacity as Secretary of Interior; U.S. DEPARTMENT OF TRANSPORTATION; ELAINE L. CHAO, in her official capacity as Secretary of Transportation; U.S. DEPARTMENT OF AGRICULTURE; SONNY PERDUE, in his official capacity as Secretary of Agriculture; U.S. DEPARTMENT OF COMMERCE; WILBUR ROSS, in his official capacity as Secretary of Commerce; U.S. DEPARTMENT OF DEFENSE; MARK T. ESPER, in his official capacity as Secretary of Defense; U.S. DEPARTMENT OF STATE; MICHAEL R. POMPEO, in his official capacity as Secretary of State; U.S. ENVIRONMENTAL PROTECTION AGENCY; ANDREW WHEELER, in his official capacity as Administrator of the EPA. These are the defendants as of September 2020.

The term "federal government" means two of the three distinct branches (legislative and executive) whose powers are vested by the US Constitution in the Congress and the President, respectively. The term also includes the hundreds of federal agencies and commissions charged with handling such responsibilities as protecting America's environment.

CHAPTER 1

1. Rob Monroe, "The History of the Keeling Curve," April 3, 2013, https://sioweb.ucsd.edu/programs/keelingcurve/2013/04/03/the-history-of-the-keeling-curve. Ex. E-1.
2. *National Science Foundation, Report on International Geophysical Year, Hearings before the Subcommittee of the Committee on Appropriations*, 85th Congress (May 1, 1957).
3. The White House, *Restoring the Quality of Our Environment: Report of the Environmental Pollution Panel, President's Science Advisory Committee*, Washington, DC, November 1965. Ex. E-289.
4. Office of Science and Technology, "Responses of the Federal Departments and Agencies to the President's Science Advisory Committee Report, 'Restoring the Quality of Our Environment,'" May 1967, 6. Ex. E-2.
5. National Science Foundation, *Weather and Climate Modification: Report of the Special Commission on Weather Modification*, 1965. Ex. E-3.
6. National Science Foundation, *Weather and Climate Modification*, 42.
7. National Science Foundation, 42.
8. National Academy of Sciences, *Weather and Climate Modification Problems and Prospects, Volume I: Summary and Recommendations*, Final Report of the Panel on Weather and Climate Modification to the Committee on Atmospheric Sciences (Washington, DC: National Academy of Sciences–National Research Council, 1966), 8. Ex. E-4.
9. US Department of Commerce, *The Automobile and Air Pollution: A Program for Progress*, Report of the Panel on Electrically Powered Vehicles to the Commerce Technical Advisory Board, Washington, DC, 1967, 5. Ex. E-5.
10. Energy Policy and Conservation Act of 1975, Pub. L. 94–163, 89 Stat. 871, Sec. 502(a)(1), enacted December 22, 1975 (setting CAFE standards at 18 mpg for model 1978 passenger cars and 27.5 mpg for model 1985 passenger cars). Ex. E-6.
11. 42 U.S.C. §§4342, 4343.
12. Daniel P. Moynihan, Memorandum to John Ehrlichman, September 17, 1969. Ex. E-7.

13. Eugene K. Peterson, "Carbon Dioxide Affects Global Ecology," *Environ. Sci. & Technology*, 3, no. 11 (1969): 1168.
14. President Richard Nixon, Letter to Lee DuBridge, Director, Office of Science and Technology, November 20, 1969. Ex. E-8.
15. Lee DuBridge, Memorandum to John Ehrlichman, Recommendation to Pursue Nonconventional Vehicle Development, December 19, 1969. Ex. E-9.
16. Council on Environmental Quality, *Environmental Quality: The First Annual Report of the Council on Environmental Quality Together With the President's Message to Congress*, August 1970. Ex. E-10.
17. Council on Environmental Quality, *First Annual Report*, 97.
18. President Richard Nixon, Annual Message to the Congress on the State of the Union, January 22, 1970 (emphasis added). Ex. E-11.
19. US Department of Commerce et al., *World Weather Program: Plan for Fiscal Year 1971*, April 1970, iv (emphasis added). Ex. E-12.
20. Edward E. David, Jr., Memorandum for Peter Flanigan, October 20, 1970. Ex. E-13. See also Sam Roberts, "Edward E. David Jr., Who Elevated Science Under Nixon, Dies at 92," *New York Times*, February 28, 2017, https://www.nytimes.com/2017/02/28/science/edward-david-dead-science-adviser-to-nixon.html. Ex. E-14.
21. Office of Science and Technology, Energy Policy Staff, Funding Energy Research and Development, August 26, 1970, 1. Ex. E-291.
22. Edward F. Hammel, "Los Alamos Scientific Laboratory Energy-Related History, Research, Managerial Reorganization Proposals, Actions Taken, and Results, 1945–1979," March 1997, 12. http://www.iaea.org/inis/collection/NCLCollectionStore/_Public/28/077/28077313.pdf. Ex. E-15.

CHAPTER 2

1. President Jimmy Carter, Second Environmental Decade Remarks at the 10th Anniversary Observance of the National Environmental Policy Act, Earth Day, and Several Federal Agencies, February 29, 1980. Ex. E-20.
2. Carter, Second Environmental Decade Remarks.
3. Carter, Second Environmental Decade Remarks.
4. President Jimmy Carter, Address to the Nation on Energy and National Goals: "The Malaise Speech," July 15, 1979. Ex. E-21.
5. The Department of Energy became an official executive agency in August 1977. Department of Energy Organization Act of 1977, Pub. L. 95-91, 91 Stat. 565, enacted August 4, 1977. The report from this early workshop was ultimately published by the DOE in May 1979.

6. US Department of Energy, *Carbon Dioxide Effects Research and Assessment Program: Workshop on the Global Effects of Carbon Dioxide from Fossil Fuels*, May 1979, v. Ex. E-22.

7. Frank Press, Memorandum to the President, Release of Fossil CO_2 and the Possibility of a Catastrophic Climate Change, July 7, 1977 (emphasis added). Ex. E-23.

8. Gordon MacDonald et al., *The Long Term Impact of Atmospheric Carbon Dioxide on Climate. Technical Report JSR-78–07*, SRI International, April 1979, 3, 24. Ex. E-24.

9. National Oceanic and Atmospheric Administration, Trends in Atmospheric Carbon Dioxide, Mauna Loa CO_2 Annual Mean Data, ftp://aftp.cmdl.noaa.gov/products/trends/co2/co2_annmean_mlo.txt. Ex. E-25.

10. Climate Research Board, National Research Council, *Carbon Dioxide and Climate: A Scientific Assessment* (Washington, DC: National Academy of Sciences, 1979), 16–17. Ex. E-27.

11. Climate Research Board, *Carbon Dioxide and Climate*, viii (emphasis added).

12. US Department of Energy, *Summary of the Carbon Dioxide Effects Research and Assessment Program*, July 1980, 1 (emphasis added). Ex. E-28.

13. Elmer B. Staats, Comptroller General, *U.S. Coal Development—Promises, Uncertainties*, EMD-77–43, Report to Congress, September 22, 1977, 6.19. Ex. E-29.

14. 15 U.S.C. §2902.

15. Council on Environmental Quality, *Environmental Quality 1977: The Eighth Annual Report of the Council on Environmental Quality*, December 1977, 188–190, 192. Ex. E-30.

16. Council on Environmental Quality, *Environmental Quality: The Ninth Annual Report of the Council on Environmental Quality*, Washington, DC, December 1978, 86. Ex. E-31.

17. Council on Environmental Quality, *Environmental Quality 1980: The Eleventh Annual Report of the Council on Environmental Quality*, Washington, DC, December 1980, 8, 12, 265–266. Ex. E-32.

18. National Research Council, *Energy and Climate: Studies in Geophysics* (Washington, DC: National Academy of Science, 1977), viii. Ex. E-33.

19. National Research Council, *Energy and Climate*, viii.

20. National Research Council, *Energy and Climate*, ix (emphasis added).

21. See, e.g., James Hansen et al., "Target Atmospheric CO_2: Where Should Humanity Aim?," *Open Atmospheric Science Journal* 2 (2008): 217–231.

22. Press, Memorandum to the President.

23. Press, Memorandum to the President.

24. Solar Energy Research, Development, and Demonstration Act of 1974, Pub. L. 93–473, 42 U.S.C. §5551 (1974).

25. Solar Energy Research, Development, and Demonstration Act.

26. Solar Energy Research, Development, and Demonstration Act.

27. President Jimmy Carter, National Energy Program Fact Sheet on the President's Program, April 20, 1977. Ex. E-34.
28. Auto fuel economy standards and building energy standards were enacted earlier in 1975.
29. Council on Environmental Quality, *Solar Energy: Progress and Promise*, Washington, DC, April 1978. Ex. E-35.
30. Council on Environmental Quality, *Solar Energy*, iii.
31. Council on Environmental Quality, 2.
32. Council on Environmental Quality, iv.
33. President Jimmy Carter, Golden, Colorado Remarks at the Solar Energy Research Institute on South Table Mountain, May 3, 1978. Ex. E-37
34. Carter, Golden, Colorado Remarks.
35. President Jimmy Carter, Solar Energy Message to the Congress, June 20, 1979. Ex. E-38.
36. Carter, Solar Energy Message.
37. Council on Environmental Quality, *The Good News About Energy*, Washington, DC, 1979, v. Ex. E-39.
38. Council on Environmental Quality, *Good News About Energy*, 27.
39. George M. Woodwell, Gordon J. MacDonald, Roger Revelle, and C. David Keeling, *The Carbon Dioxide Problem: Implications for Policy in the Management of Energy and Other Resources*, A Report to the Council on Environmental Quality, July 1979, reprint 2008. Ex. E-40.
40. Woodwell et al., *Carbon Dioxide Problem*, 7–8.
41. Woodwell et al., 7, 11, 12, 13.
42. Our 1979 CEQ report echoed these recommendations, saying: "Conceivably, scientific proof of the warming of the earth might come after the time has passed when action could be taken to reverse the trend. 'If we wait until there is absolute proof that the increase in CO_2 is causing a warming of the earth,' says Dr. George Woodwell of Woods Hole Marine Biological Laboratory, 'it will be 20 years too late to do anything about it.'" Council on Environmental Quality, *Environmental Quality: The Tenth Annual Report of the Council on Environmental Quality*, Washington, DC, 1979, 622. Ex. E-41.
43. Philip Shabecoff, "Scientists Warn U.S. of Carbon Dioxide Peril," *New York Times*, July 10, 1979. Ex. E-26.
44. Shabecoff, "Scientists Warn U.S."
45. A different 1980 report by the National Academy of Sciences, conveyed to President Carter in a memorandum by Frank Press, made clear that the Carter administration understood the connection between energy policy and climate change. The 1980 report noted that "preventing or delaying the increase in carbon dioxide would have to be done mainly by

restricting the use of fossil fuels, although management of land and forests could also contribute." Also, "increases in energy consumption using fossil fuels will have increasingly undesirable climate effects. . . . We and the main energy-consuming countries must keep open a number of options for energy and not become committed to an extended period of unrestricted fossil-fuel use." Frank Press, Memorandum to the President, Carbon Dioxide Increases, May 5, 1980. Ex. E-42.

46. Gerald O. Barney, ed., *Global 2000: The Report to the President: Entering the Twenty-First Century*, 1980, reissued in 1988, 3. Ex. E-43.

47. Barney, *Global 2000*, 5.

48. Council on Environmental Quality and US Department of State, *Global Future: Time to Act*, Washington, DC, January 1981, 63, 138. Ex. E-44.

49. Philip Shabecoff, "U.S. Calls for Efforts To Combat Global Environmental Problems," *New York Times*, January 15, 1981. Ex. E-45.

50. Council on Environmental Quality, *Global Energy Futures and the Carbon Dioxide Problem*, Washington, DC, January 1981, 65–73. Ex. E-46.

51. Council on Environmental Quality, *Global Energy Futures*, vi.

52. Philip Shabecoff, "U.S. Study Warns of Extensive Problems from Carbon Dioxide Pollution," *New York Times*, January 14, 1981. Ex. E-47.

53. Carter, National Energy Program Fact Sheet.

54. President Jimmy Carter, Outer Continental Shelf Lands Act Amendments of 1978 Statement on Signing S.9 Into Law, September 18, 1978. Ex. E-48.

55. President Jimmy Carter, The State of the Union Annual Message to the Congress, January 16, 1981. Ex. E-49.

56. Carter, State of the Union 1981.

57. Carter.

58. John B. Oakes, "For Reagan, a Ticking Ecological 'Time Bomb,'" *New York Times*, January 20, 1981. Ex. E-50.

59. Oakes, "Ticking Ecological 'Time Bomb.'"

60. Oakes.

61. US Energy Information Administration, *Monthly Energy Review August 2020*, 2020, table 1.3.

CHAPTER 3

1. David Biello, "Where Did the Carter White House's Solar Panels Go?," *Scientific American*, August 6, 2010.

2. US Department of Energy, *Carbon Dioxide Effects Research and Assessment Program: Proceedings of the Workshop on First Detection of Carbon Dioxide Effects*, Harpers Ferry, West Virginia, June 8–10, 1981, Washington, DC, May 1982. Ex. E-52.

3. US Department of Energy, *Proceedings of the Workshop*, vii.

4. US Department of Energy, 3.

5. US Department of Energy, *Carbon Dioxide Effects Research and Assessment Program: Environmental and Societal Consequences of a Possible CO_2-Induced Climate Change*, Vol. II, Part I, Response of the West Antarctic Ice Sheet to CO_2-Induced Climatic Warming, Washington, DC, April 1982, 1. Ex. E-53.

6. David Narum, "A Troublesome Legacy: The Reagan Administration's Conservation and Renewable Energy Policy," *Energy Policy* 20, no. 1 (1992). http://www.sciencedirect.com/science/article/pii/030142159290146S.

7. National Research Council, *Carbon Dioxide and Climate: A Second Assessment* (Washington, DC: National Academy Press, 1982). Ex. E-54.

8. National Research Council, *Carbon Dioxide and Climate*, 1 (emphasis added).

9. National Research Council, 14.

10. National Research Council, *Changing Climate: Report of the Carbon Dioxide Assessment Committee* (Washington, DC: National Academy Press, 1983), 8. https://doi.org/10.17226/18714.

11. US Environmental Protection Agency, *Can We Delay a Greenhouse Warming?*, Washington, DC, September 1983, iv, 1–7 to 1–8. Ex. E-55.

12. US Environmental Protection Agency, *Can We Delay*, i.

13. US Environmental Protection Agency, 1–16 to 1–17.

14. US Environmental Protection Agency, 1–11.

15. US Environmental Protection Agency, *Projecting Future Sea Level Rise: Methodology, Estimates to the Year 2100, and Research Needs*, Washington, DC, October 1983, vi. Ex. E-56.

16. US Environmental Protection Agency, *Projecting Future Sea Level*, 8.

17. US Environmental Protection Agency, 8.

18. William D. Ruckelshaus, Remarks at Organization for Economic Cooperation and Development, June 21, 1984, 6–7. Ex. E-57.

19. Ruckelshaus, Remarks, 5.

20. William D. Ruckelshaus, The Role of the Private Sector in Environmental Action, World Industry Conference on Environmental Management, November, 14, 1984, 4, 24. Ex. E-58.

21. US Department of Energy, John R. Trabalka, ed., *Atmospheric Carbon Dioxide and the Global Carbon Cycle*, Washington, DC, December 1985, xvi. Ex. E-59. The other volume

was "Direct Effects of Increasing Carbon Dioxide on Vegetation," with two companion reports also published ("Characterization of Information Requirements for Studies of CO_2 Effects: Water Resources, Agriculture, Fisheries, Forests, and Human Health" and "Glaciers, Ice Sheets, and Sea Level: Effects of a CO_2-Induced Climatic Change").

22. US Department of Energy, Michael C. MacCracken and Frederick M. Luther, eds., *Projecting the Climatic Effects of Increasing Carbon Dioxide*, Washington, DC, December 1985, ix. Ex. E-60.

23. US Department of Energy, *Projecting the Climatic Effects*, 241.

24. US Department of Energy, 271–272.

25. National Oceanic and Atmospheric Administration, "Global Climate: A Variable and Vulnerable Natural Resource," October 1, 1986, 1. Ex. E-61.

26. National Oceanic and Atmospheric Administration, "Global Climate," 36 (emphasis added).

27. Senators George J. Mitchell et al., Letter to Lee Thomas, Administrator, Environmental Protection Agency, September 12, 1986. Ex. E-62.

28. Mitchell et al., Letter to Thomas (emphasis added).

29. Senators Robert T. Stafford et al., Letter to John Gibbons, Executive Director, US Congress Office of Technology Assessment, December 23, 1986. Ex. E-292.

30. US Environmental Protection Agency, "Climate Change Indicators: Climate Forcing," https://www.epa.gov/climate-indicators/climate-change-indicators-climate-forcing. Ex. E-63.

31. US Department of Energy, Michael P. Farrell, ed., *Master Index for the Carbon Dioxide Research State-of-the-Art Report Series*, Washington, DC, March 1987. Ex. E-64.

32. *Carbon Dioxide and Climate: The Greenhouse Effect, Hearing before the Subcommittee on Natural Resources, Agricultural Research and Environment and the Subcommittee on Investigations and Oversight of the Committee on Science and Technology*, 97th Congress 39 (March 25, 1982). Ex. E-65.

33. *Carbon Dioxide and Climate*, 48.

34. Dotty Curlee, Department of Energy Action Officer, Report of Hearing, Joint Hearing on Carbon Dioxide and the Greenhouse Effect, March 25, 1982 (emphasis in original). Ex. E-66.

35. *Carbon Dioxide and Climate*, 12.

36. *Ozone Depletion, the Greenhouse Effect, and Climate Change, Hearings before the Subcommittee on Environmental Pollution of the Committee on Environment and Public Works*, 99th Congress 22 (June 10 and 11, 1986). Ex. E-67.

37. *Ozone Depletion*, 97.

38. *Greenhouse Effect and Global Climate Change, Hearing before the Committee on Energy and Natural Resources*, 100th Congress 39 (June 23, 1988). Ex. E-68.

39. *Greenhouse Effect and Global Climate Change*, 93–94.

40. Philip Shabecoff, "Global Warming Has Begun, Expert Tells Senate," *New York Times*, June 24, 1988. Ex. E-69.

41. *Energy Policy Implications of Global Warming, Hearings before the Subcommittee on Energy and Power of the Committee on Energy and Commerce*, 100th Congress 85, 100 (July 7 and September 22, 1988). Ex. E-70.

42. *Energy Policy Implications*, 192.

43. Council on Environmental Quality, *Environmental Quality 1981: 12th Annual Report of the Council on Environmental Quality*, Washington, DC, 1982, 191. Ex. E-71.

44. Council on Environmental Quality, *Environmental Quality 1983: 14th Annual Report of The Council on Environmental Quality*, Washington, DC. Ex. E-72; Council on Environmental Quality, *Environmental Quality: 15th Annual Report of The Council on Environmental Quality*, Washington DC, 714. Ex. E-290.

45. Al Gore, Letter to The Honorable Ronald W. Reagan, May 21, 1986. Ex. E-73.

46. William L. Ball, Assistant to the President, Letter to Senator Gore, May 28, 1986. Ex. E-73.

47. President Ronald Reagan, Message to the Congress Transmitting the National Energy Policy Plan, July 17, 1981. Ex. E-74.

48. Council on Environmental Quality, *12th Annual Report*, 148 (quoting 1981 National Energy Policy Plan).

49. Council on Environmental Quality, 149.

50. Council on Environmental Quality, *Environmental Quality 1982: 13th Annual Report of The Council on Environmental Quality*, Washington, DC, 156. Ex. E-75.

51. Council on Environmental Quality, *13th Annual Report*, 156.

52. Rolando A. Gächter, *Federal Offshore Statistics: 1995: Leasing, Exploration, Production, and Revenue as of December 31, 1995*, Herdon, VA: US Department of Interior, 1997, 6. Ex. E-76.

53. Gächter, *Federal Offshore Statistics: 1995*, 6.

54. Council on Environmental Quality, *13th Annual Report*, 160.

55. See, e.g., John S. Herrington, DOE Secretary, Letter to The President, March 16, 1987. Ex. E-77 (describing an impending oil shortage and what "has already been done during this Administration to strengthen the domestic oil industry and remove impediments to the exploration for oil and gas").

56. See, e.g., George Cameron Coggins and Doris K. Nagel, "'Nothing Beside Remains': The Legal Legacy of James G. Watt's Tenure as a Secretary of the Interior on Federal Land and Law Policy," *Boston College Environmental Affairs Law Review* 17, no. 3 (1990): 521–522 (describing Watt's efforts to accelerate leasing of oil and gas from public lands "in the face of powerful reasons to proceed more cautiously").

57. US Environmental Protection Agency, Regulatory Determination for Oil and Gas and Geothermal Exploration, Development and Production Wastes, 53 Fed. Reg. 25446 (July 6, 1988). Ex. E-78.

58. Office of Science and Technology Policy, *Biennial Science and Technology Report to Congress: 1983–1984*, Washington, DC, 1985, 79–81. Ex. E-79.

59. US Department of Transportation, *Summary of Fuel Economy Performance*, Washington, DC, December 15, 2014. Ex. E-80; The PEW Charitable Trusts, *Driving to 54.5 mpg: A History of Fuel Efficiency in the United States*, September 2012, 3. http://www.pewtrusts.org/-/media/assets/2014/06/02/factsheet-graphic-fuel-effiency-timeline-finalsept-2012.pdf. Ex. E-81.

60. World Climate Programme, *Report of the International Conference on the Assessment of the Role of Carbon Dioxide and of Other Greenhouse Gases in Climate Variations and Associated Impacts*, Villach, Austria, 9–15 October 1985, 1986, 1. Ex. E-82.

61. United Nations, *Our Common Future: Report of the World Commission on Environment and Development*, 1987. Ex. E-83.

62. World Meteorological Organization, "Conference Proceedings: The Changing Atmosphere: Implications for Global Security, Toronto, Canada, 27–30 June 1988," https://wedocs.unep.org/handle/20.500.11822/29980.

63. Richard J. Smith, Acting Assistant Secretary of State for Oceans and International Environmental and Scientific Affairs, Memorandum to Dr. Ralph Bledsoe, White House Domestic Policy Council, January 15, 1988. Ex. E-84 (discussing President's obligation under the Global Climate Protection Act to develop a national policy on climate change).

64. J. Edward Fox, Assistant Secretary Legislative Affairs, Letter to Senator Chafee, July 29, 1988. Ex. E-85; see also Richard J. Smith, Acting Assistant Secretary of State for Oceans and International Environmental and Scientific Affairs, Letter to Richard Hallgren, January 27, 1988. Ex. E-86.

65. Global Climate Protection Act of 1987.

66. Robert E. Johnson, Memorandum through Ralph C. Bledsoe for Nancy J. Risque, The Global Climate Protection Act, December 29, 1987. Ex. E-87.

67. President Ronald Reagan, Report to Congress: United States Government Activities Related to the Greenhouse Effect, January 26, 1988. Ex. E-88.

CHAPTER 4

1. See "The White House and the Greenhouse," *New York Times*, May 9, 1989. Ex. E-89.

2. US Department of State, Remarks by the Honorable James A. Baker III Secretary of State before the Response Strategies Working Group, Intergovernmental Panel on Climate Change, January 30, 1989. Ex. E-90.

3. Frederick M. Bernthal, Memorandum to Richard T. McCormack, Under Secretary-Designate for Economic Affairs, Attached Background Material, February 9, 1989. Ex. E-91.
4. Frederick M. Bernthal, Memorandum to the Secretary of the Department of State, Review of Key Foreign Policy Issues: The Environment, February 27, 1989. Ex. E-92.
5. US Environmental Protection Agency, *The Potential Effects of Global Climate Change on the United States*, Washington, DC, December 1989. Ex. E-93.
6. US Environmental Protection Agency, *Potential Effects*, 9.
7. US Environmental Protection Agency, xxxiv.
8. US Environmental Protection Agency, xxxii.
9. Council on Environmental Quality, *Environmental Quality: The Twentieth Annual Report of the Council on Environmental Quality together with The President's Message to Congress*, Washington, DC, 1990, 227, 278. Ex. E-94.
10. Council on Environmental Quality, *Environmental Quality: The Twenty-Third Annual Report of the Council on Environmental Quality together with The President's Message to Congress*, Washington, DC, January 1993, 144. Ex. E-95. US Environmental Protection Agency, *Inventory of U.S. Greenhouse Gas Emissions and Sinks: 1990–2016*, 2018. Ex. E-96.
11. US General Accounting Office, *Greenhouse Effect: DOE's Programs and Activities Relevant to the Global Warming Phenomenon*, Washington, DC, March 1990, 14. Ex. E-97.
12. US General Accounting Office, *Global Warming: Emission Reductions Possible as Scientific Uncertainties are Resolved*, Washington, DC, September 1990, 4. Ex. E-98.
13. US General Accounting Office, *Global Warming: Emission Reductions*, 4.
14. National Oceanic and Atmospheric Administration, *Reports to the Nation on Our Changing Planet: The Climate System*, Winter 1991. Ex. E-99
15. National Oceanic and Atmospheric Administration, *Reports to the Nation*, 11.
16. National Oceanic and Atmospheric Administration, 17–18.
17. 15 U.S.C. §2931(a)(1)–(2).
18. Bernthal, Memorandum to McCormack.
19. Bernthal.
20. Bernthal, Memorandum to the Secretary, 2.
21. Solar Energy Research Institute, *CO_2 Emissions from Coal-Fired and Solar Electric Power Plants*, May 1990. Ex. E-100.
22. Solar Energy Research Institute, *CO_2 Emissions from Coal-Fired*, 3.
23. Solar Energy Research Institute, 28.

24. US Department of Energy, *The Economics of Long-Term Global Climate Change: A Preliminary Assessment*, Report of an Interagency Task Force, Washington, DC, September 1990. Ex. E-101.

25. US Department of Energy, *Economics of Long-Term*, 29, 31.

26. US General Accounting Office, *Energy Policy: Options to Reduce Environmental and Other Costs of Gasoline Consumption*, Washington, DC, September 1992, 2. Ex. E-102.

27. US Environmental Protection Agency, *Policy Options for Stabilizing Global Climate: Report to Congress*, December 1990, xxiii. Ex. E-103.

28. US Environmental Protection Agency, *Policy Options for Stabilizing Global Climate*, 1, 2.

29. See, e.g., Senators George J. Mitchell et al., Letter to Lee Thomas, Administrator, Environmental Protection Agency, September 12, 1986. Ex. E-62 (asking the EPA to consider policy options that would stabilize atmospheric GHG emissions at current levels, which were right around 350 ppm).

30. US Environmental Protection Agency, *Policy Options for Stabilizing Global Climate*, 8, VI-47.

31. US Environmental Protection Agency, VI-1.

32. US Environmental Protection Agency, V-1 (emphasis added).

33. US Environmental Protection Agency, I-13.

34. US Congress, Office of Technology Assessment, *Changing by Degrees: Steps To Reduce Greenhouse Gases, OTA-O-482* (Washington, DC: US Government Printing Office, February 1991), Foreword. Ex. E-104.

35. Office of Technology Assessment, *Changing by Degrees*, 5.

36. Office of Technology Assessment, 12–13.

37. Office of Technology Assessment, 5.

38. Office of Technology Assessment, 3.

39. Alan Hill, Chairman, Council on Environmental Quality, Memorandum to Heads of Federal Agencies, Draft of Guidance to Federal Agencies Regarding Consideration of Global Climate Change in Preparation of Environmental Documents, June 21, 1989. Ex. E-105.

40. See Donald A. Carr, Acting Assistant Attorney General, Memorandum to Kenneth Yale, CEQ Global Climate Change Directive, June 23, 1989. Ex. E-106.

41. Carr, Memorandum to Yale, 1.

42. Carr, 2–3.

43. David Bates, Assistant to the President and Secretary to the Cabinet, Draft Memorandum to Heads of Agencies, July 14, 1989. Ex. E-107.

44. US General Accounting Office, *Global Warming: Administration Approach Cautious Pending Validation of Threat*, Washington, DC, January 1990, 17. Ex. E-108.

45. US General Accounting Office, *Global Warming: Administration Approach*, 3, 21.

46. *Climate Surprises, Hearing before the Subcommittee on Science, Technology, and Space of the Committee on Commerce, Science, and Transportation*, 101st Congress (May 8, 1989). Ex. E-109; see also Philip Shabecoff, "Scientist Says Budget Office Altered His Testimony," *New York Times*, May 8, 1989. Ex. E-110. James Hansen, *Storms of My Grandchildren: The Truth About the Coming Climate Catastrophe and Our Last Chance to Save Humanity* (New York: Bloomsbury USA, 2009).

47. *Climate Surprises*, 143–144, 147.

48. Schneider would later serve as a consultant to the first Bush Administration, and holds the rare distinction of having at one point served in a consulting or advisory capacity for every administration from Richard Nixon to Barack Obama. See NCAR UCAR, "Stephen Schneider: An Extraordinary Life," July 21, 2010, https://news.ucar.edu/2270/stephen-schneider-extraordinary-life. Ex. E-111.

49. Included in Briefing for Governor Sununu, Preparations for the First Framework Convention Negotiating Session, Washington, DC, February 4, 1991, November 7, 1990. Ex. E-112.

50. Patrick Michaels and Robert Balling, Jr., Letter to The Honorable George H. Bush, February 1, 1991. Ex. E-113.

51. D. Allen Bromley, Memorandum to John H. Sununu, Possible Group from Whom to Obtain Reaction to the NAS/Evans Report, April 18, 1991. Ex. E-114.

52. See Institute of Medicine, National Academy of Sciences, and National Academy of Engineering, *Policy Implications of Greenhouse Warming* (Washington, DC: National Academy Press, 1991). Ex. E-115; Bromley, Memorandum to Sununu, Possible Group; D. Allen Bromley, Memorandum to Governor Sununu, The NRC Evans Report, April 8, 1991. Ex. E-116; Nancy G. Maynard, Memorandum for D. Allen Bromley, J. Thomas Ratchford, Environment Update—Week of May 26 and June 3, 1991, June 3, 1991. Ex. E-117.

53. David W. Loer, Letter to Mr. John Sununu, Chief of Staff, July 30, 1991. Ex. E-118.

54. John Sununu, Letter to David W. Loer, August 20, 1991 (emphasis in original). Ex. E-118.

55. President George H. W. Bush, Presidential Address: Intergovernmental Panel on Climate Change, Georgetown University, February 5, 1990. Ex. E-119.

56. Intergovernmental Panel on Climate Change, *Climate Change: The IPCC Scientific Assessment* (Cambridge: Cambridge University Press, 1990).

57. See Allan Bromley and Stephen Danzansky, Memorandum for Fred Bernthal et al., Meeting of Global Change Strategy Task Force, May 21, 1990. Ex. E-120; D. Allan Bromley and Ede Holiday, Memorandum for Governor Sununu, Climate Change Framework Convention Meeting at 9:30 Wednesday morning, January 23, 1991. Ex. E-121. Policy decisions on climate change came from John Sununu. See Michael Weisskopf, "Bush Was Aloof in Warming Debate," *Washington Post*, October 31, 1992. Ex. E-122.

58. Frederick M. Bernthal, Letter to Professor Bert Bolin, July 5, 1990. Ex. E-123; Steve Danzansky, Memorandum for Chris Dawson et al., Letter on IPCC Conclusions Paper, August 7, 1990. Ex. E-124.

59. D. Allen Bromley, Memorandum for John H. Sununu, Sea Level Change and Environmental Matters, June 14, 1990. Ex. E-125.

60. Richard J. Smith, Confidential Information Memorandum to The Secretary, Department of State, Preparations for an International Conference on the Environment, May 16, 1989. Ex. E-126.

61. In US Department of State Memorandum entitled PRD-12/Global Climate Change Policy Decision Paper, the Department of State made the following statement: "While many nations sought to set firm 'targets and timetables' for reducing greenhouse gas emissions, U.S. objection to firm commitments resulted in an agreement that sets a non-binding goal for developed countries to return emissions to 1990 levels by the end of the decade." US Department of State, "PRD-12/Global Climate Change Policy Decision Paper." Ex. E-127. See also Roger B. Porter, Memorandum for the President, The Second World Climate Conference, October 23, 1990. Ex. E-128; Global Change Working Group, Memorandum for the Domestic Policy Council, Framework Convention on Climate Change, January 22, 1991, 7. Ex. E-129. Internal memos from the time period also show resistance to targets and timetables in the White House. See, e.g., Bromley and Dazansky, Memorandum for Bernthal et al. ("Targets and Timetables: We do not believe there is sufficient evidence at this time to warrant stringent measures, with potentially serious negative economic consequences, to limit greenhouse gas emissions").

62. William A. Nitze, "A Failure of Presidential Leadership," in *Negotiating Climate Change: The Inside Story of the Rio Convention* (Cambridge: Cambridge University Press, 1994), 189.

63. Nathaniel Rich, *Losing Earth, A Recent History* (New York: Farrar, Straus and Giroux, 2019), 171.

64. See also Nitze, "Failure of Presidential Leadership," 194.

65. United Nations, *United Nations Framework Convention on Climate Change*, 1992, Preamble. Ex. E-130.

66. United Nations, *United Nations Framework Convention*, Preamble.

67. United Nations, Preamble.

68. United Nations, Preamble.

69. United Nations, Article 2 (emphasis added).

70. United Nations, Article 4.

71. United Nations, Article 4.

72. United Nations, Article 4.

73. President George H. W. Bush, Remarks on Signing the Natural Gas Wellhead Decontrol Act of 1989, July 26, 1989. Ex. E-131.

74. US Department of Energy, *National Energy Strategy: Powerful Ideas for America*, Washington, DC, February 1991, 19, 74–106. Ex. E-132.

75. US Department of Energy, *National Energy Strategy*, 10–11, 74–106.

76. US Department of Energy, *National Energy Strategy*, 1–7; US General Accounting Office, *Energy Conservation: DOE's Efforts to Promote Energy Conservation and Efficiency*, Washington, DC, April 1992, 5. Ex. E-133.

77. US Department of Energy, *National Energy Strategy*, 10–11.

78. US Department of Energy, 12.

79. See US General Accounting Office, *Greenhouse Effect: DOE's Programs*, 17.

80. US General Accounting Office, 15.

81. Associated Press, "The Earth Summit; Excerpts from Speech by Bush on 'Action Plan,'" *New York Times*, June 13, 1992, https://www.nytimes.com/1992/06/13/world/the-earth-summit-excerpts-from-speech-by-bush-on-action-plan.html. Ex. E-134.

82. President George H. W. Bush, Remarks on Signing the Energy Policy Act of 1992 in Maurice, Louisiana, October 24, 1992. Ex. E-135.

83. Energy Policy Act of 1992, Pub. L. 102–486, October 24, 1992. Ex. E-136.

84. G. H. W. Bush, Remarks on Signing the Energy Policy Act.

85. US Environmental Protection Agency, *Policy Options for Stabilizing Global Climate*, V-15, V-22.

86. The PEW Environment Group, *History of Fuel Economy*, April 2011, http://www.pewtrusts.org/~/media/assets/2011/04/history-of-fuel-economy-clean-energy-factsheet.pdf. Ex. E-137.

CHAPTER 5

1. Al Gore, *Earth in the Balance: Ecology and the Human Spirit* (Boston: Houghton Mifflin, 1992).

2. President William J. Clinton, Remarks on Earth Day, April 21, 1993. Ex. E-138.

3. Clinton, Remarks on Earth Day 1993.

4. Intergovernmental Panel on Climate Change, *IPCC Second Assessment Climate Change 1995: A Report of the Intergovernmental Panel on Climate Change*, 1995.

5. Timothy E. Wirth, Undersecretary for Global Affairs on Behalf of the United States of America, Statement at the Second Conference of the Parties, Framework Convention on Climate Change, July 17, 1996. Ex. E-139.

6. Vice President Al Gore, Remarks at the SE Regional Climate Change Impacts Meeting, Vanderbilt University, Nashville, Tennessee, June 25, 1997. Ex. E-140.

7. See also US Congress, Office of Technology Assessment, *Preparing for an Uncertain Climate: Volume 1, OTA-0-567* (Washington DC: US Government Printing Office, October 1993). Ex. E-141; Office of Science and Technology Policy, *Climate Change: State of Knowledge*, 1997. Ex. E-142; National Science and Technology Council and the Institute of Medicine/National Academy of Sciences, *Conference on Human Health and Global Climate Change: Summary of the Proceedings* (Washington, DC: National Academy Press, 1996). Ex. E-143; Tom Ross and Neal Lott, "A Climatology of Recent Extreme Weather and Climate Events," National Oceanic and Atmospheric Administration, National Climatic Data Center, October 2000. Ex. E-183.

8. US Environmental Protection Agency, *Effects of CO_2 and Climate Change on Forest Trees*, Environmental Research Laboratory—Corvallis, April 1993, 22–23. Ex. E-144.

9. US Environmental Protection Agency, *Effects of CO_2 and Climate Change*, 24.

10. US Department of State, "PRD-12/Global Climate Change," 1.

11. US Department of State, 1–2.

12. Katie McGinty, Memorandum to the President and the Vice President, Climate Change Action Plan, October 8, 1993. Ex. E-145.

13. J. D. Mahlman, D. Albritton, and R. T. Watson, State of Scientific Understanding of Climate Change, included in Katie McGinty, Memorandum to the President and the Vice President, Climate Change Action Plan, October 8, 1993. Ex. E-146.

14. President William J. Clinton, Remarks at the White House Conference on Climate Change, October 19, 1993. Ex. E-147.

15. Council on Environmental Quality, *Environmental Quality: Along the American River: The 1996 Report of the Council on Environmental Quality*, Washington, DC, xi. Ex. E-148.

16. Council on Environmental Quality, *Environmental Quality: The 1997 Report of the Council on Environmental Quality*, Washington, DC, 163, 194. Ex. E-149.

17. Intergovernmental Panel on Climate Change, *IPCC Second Assessment Climate Change 1995*, 3.

18. The White House, Office of the Press Secretary, Remarks by the President on Global Climate Change, National Geographic Society, Washington, DC, October 22, 1997. Ex. E-150. See also Katie McGinty, Dan Albritton, and Jerry Melillo, Press Briefing, Climate Change Briefing, July 24, 1997. Ex. E-151 (Dan Albritton: "This is not the view of one scientist, this is the view of thousands of scientists who were asked to give their current statement of scientific understanding. I believe that those statements made by the entire scientific community are of high value to decision-makers in government and industry and those who acquaint the public with such complex things").

19. Clinton, Remarks on Earth Day 1993.

20. President William J. Clinton, Remarks on the 25th Observance of Earth Day in Havre de Grace, Maryland, April 21, 1995. Ex. E-152.

21. The White House, Office of the Press Secretary, Remarks by the President to Business Roundtable, Washington, DC, June 12, 1997. Ex. E-153.

22. The White House, Office of the Press Secretary, Opening Remarks by the President and the Vice President at Discussion on Climate Change, July 24, 1997. Ex. E-154.

23. The White House, Office of the Press Secretary, Remarks by the President at White House Conference on Climate Change, Georgetown University, October 6, 1997, 4. Ex. E-155.

24. The White House, Opening Remarks by the President and the Vice President.

25. US Environmental Protection Agency, *Environmental Health Threats to Children*, Washington, DC, September 1996. Ex. E-156.

26. Exec. Order No. 13045, Protection of Children From Environmental Health Risks and Safety Risks, 62 Fed. Reg. 19885 (April 23, 1997). Ex. E-157.

27. Timothy E. Wirth, Undersecretary of State, Remarks at the First Conference of the Parties to the Framework Convention on Climate Change, April 5, 1995. Ex. E-158.

28. Council on Environmental Quality, *Along the American River*, xi.

29. National Science and Technology Council, *Conference on Human Health*, 9.

30. Wirth, Remarks at the First Conference of the Parties.

31. Wirth, Statement at the Second Conference of the Parties.

32. National Science and Technology Council, *Conference on Human Health*, 3–4 (Al Gore, The Interplay of Climate Change, Ozone Depletion, and Human Health, excerpts from Vice President Al Gore's remarks at the Conference on Human Health and Global Climate Change, September 11, 1995).

33. US Environmental Protection Agency, *Effects of* CO_2 *and Climate Change*, 17–18 (internal citations omitted).

34. Linda A. Joyce and Richard Birdsey, technical eds., *The Impact of Climate Change on America's Forests: A Technical Document Supporting the 2000 USDA Forest Service RPA Assessment* (Fort Collins, CO: US Department of Agriculture, Forest Service, Rocky Mountain Research Station, 2000). Ex. E-159.

35. Joyce and Birdsey, *Impact of Climate Change on America's Forests*, 34, 128 (internal quotations omitted).

36. US Geological Survey, "Coastal Wetlands and Global Change: Overview," June 1997, 4. Ex. E-160.

37. US Geological Survey, "Coastal Wetlands," 1.

38. US Department of the Interior, *Climatic Change in the National Parks, Wildlife Refuges and Other Department of Interior Lands in the United States*, May 1997. Ex. E-161.

39. US Department of the Interior, *Climatic Change in the National Parks*, 1.

40. See, e.g., McGinty, Albritton, and Melillo, Press Briefing. (Jerry Melillo of OSTP discussing melting permafrost in Alaska from climate change and that this could make it difficult to keep rural airports functioning and roads flat—two essential ways that people move around Alaska); The White House, "Global Climate Change: An East Room Roundtable," July 24, 1997, 4–5. Ex. E-162 (a White House roundtable meeting discussing loss of sugar maples in New England Forests, submergence of salt marshes in Louisiana, and loss of glacial areas in Glacier National Park); Gene Sperling, Katie McGinty, and Daniel Tarullo, Memorandum for the President, Climate Change Scenarios, September 15, 1997, 16. Ex. E-163 (a memorandum to President Clinton discussing "a rise in sea levels that will inundate more than 9000 square miles in the United States (with Florida and Louisiana most vulnerable)," from a temperature increase of 2–6.5°F above preindustrial levels).

41. United Nations, *United Nations Framework Convention*, Article 2; United Nations, *Kyoto Protocol to the United Nations Framework Convention on Climate Change*, 1998. Ex. E-164.

42. John H. Cushman, Jr., "U.S. Signs a Pact to Reduce Gases Tied to Warming," *New York Times*, November 13, 1998, https://www.nytimes.com/1998/11/13/world/us-signs-a-pact-to-reduce-gases-tied-to-warming.html. Ex. E-165.

43. United Nations, *Kyoto Protocol.*

44. United Nations.

45. S.Res. 98, A resolution expressing the sense of the Senate regarding the conditions for the United States becoming a signatory to any international agreement on greenhouse gas emissions under the United Nations Framework Convention on Climate Change, 105th Congress (1997).

46. Cushman, "U.S. Signs a Pact."

47. See The White House, Remarks by the President at White House Conference; US General Accounting Office, *Global Warming: Administration's Proposal in Support of the Kyoto Protocol*, Statement of Victor S. Rezendes, Testimony Before the Committee on Energy and Natural Resources, June 4, 1998. Ex. E-166.

48. Jon A. Krosnick et al., "American Opinion on Global Warming: The Impact of the Fall 1997 Debate," *Resources* 133 (1998): 8. Ex. E-167.

49. President William J. Clinton and Vice President Al Gore, Jr., *The Climate Change Action Plan* (Washington, DC: Executive Office of the President, 1993), 7. Ex. 168. But see Q&A document on the Plan, cautioning that "this plan by itself is unlikely to stabilize emissions at 1990 levels under reasonable assumptions regarding economic growth, the diffusion of existing technologies, and new technology development." Selected Questions and Answers on the President's Climate Change Action Plan, in Kathleen McGinty, Letter to Colleague, October 18, 1993. Ex. E-169.

50. Congressional Research Service, *U.S. Greenhouse Gas Emissions: Recent Trends and Factors*, November 24, 2014. Ex. E-170.

51. See Clinton and Gore, *Climate Change Action Plan*, 20. (The plan directed EPA, DOE, and FERC to encourage the use of natural gas.)

52. US Department of Energy, Office of Fossil Energy, *Environmental Benefits of Advanced Oil and Gas Exploration and Production Technology*, Washington, DC, October 1999, 8. Ex. E-172.

53. Domestic Policy Council, Domestic Gas and Oil Incentives, 1994, 15, 34. Ex. E-173.

54. Deep Water Royalty Relief Act, Pub. L. 104–58, November 28, 1995.

55. See also Rick Morgan, Fax to Mike Toman, Post-2000 Option: CO_2, Cap & Emissions Trading for Electricity, Natural Gas, and Transportation Fuels, October 21, 1994. Ex. E-174.

56. Jonathan Z. Cannon, Memorandum to Carol M. Browner, EPA's Authority to Regulate Pollutants Emitted by Electric Power Generation Sources, April 10, 1998. Ex. E-175.

57. Selected Questions and Answers on the President's Climate Change Action Plan.

58. US Department of Energy, *Million Solar Roofs: Become One In A Million*, Washington, DC, 2003. Ex. E-176.

59. Solar Energy Industries Association, "Solar Industry Research Data," accessed August 3, 2018, https://www.seia.org/solar-industry-research-data. Ex. E-177.

60. L. T. Flowers and P. J. Dougherty, "Wind Powering America: Goals, Approach, Perspectives, and Prospects," Preprint, National Renewable Energy Laboratory, March 2002, 1. Ex. E-178.

CHAPTER 6

1. Committee on Oversight and Government Reform, US House of Representatives, *Political Interference with Climate Change Science Under the Bush Administration*, December 2007, i. Ex. E-179.

2. Talking Points for Governor Christine Todd Whitman, Administrator, United States Environmental Protection Agency at the G8 Environmental Ministerial Meeting Working Session on Climate Change, Trieste, Italy, March 3, 2001. Ex. E-180.

3. Intergovernmental Panel on Climate Change, *Climate Change 2001: The Scientific Basis. Contribution of Working Group I to the Third Assessment Report of the Intergovernmental Panel on Climate Change* (Cambridge: Cambridge University Press, 2001), 10.

4. Intergovernmental Panel on Climate Change, *Climate Change 2001*, 13.

5. National Research Council, *Climate Change Science: An Analysis of Some Key Questions* (Washington, DC: National Academy Press 2001), Appendix A. Ex. E-181.

6. National Research Council, *Climate Change Science*, 1.

7. National Research Council, 1.

8. US Department of State, *U.S. Climate Action Report 2002: Third National Communication of the United States of America Under the United Nations Framework Convention on Climate Change*, Washington, DC, May 2002, 4–5. Ex. E-182.

9. US Department of State, *U.S. Climate Action Report 2002*, 82, 84, 103.

10. US Department of State, 84.

11. US Department of State, 6.

12. National Oceanic and Atmospheric Administration National Centers for Environmental Information, "State of the Climate: Hurricanes and Tropical Storms—Annual 2005," January 2006, accessed on July 26, 2016, https://www.ncdc.noaa.gov/sotc/tropical-cyclones/200513. Ex. E-184.

13. Paul Marshall and Heidi Schuttenberg, *A Reef Manager's Guide to Coral Bleaching* (Great Barrier Reef Marine Park Authority, 2006). Ex. E-185.

14. J. Richter-Menge et al., *State of the Arctic Report*, National Oceanic and Atmospheric Administration, October 2006. Ex. E-186.

15. Marshall and Schuttenberg, *Reef Manager's Guide*, 109–110.

16. *Climate Change: Understanding the Degree of the Problem, Hearing before the Committee on Government Reform*, 109th Congress 88 (July 20, 2006). Ex. E-187.

17. Richter-Menge et al., *State of the Arctic Report*, 24.

18. Richter-Menge et al., 28 (emphasis added).

19. Congressional Budget Office, *Evaluating the Role of Prices and R&D in Reducing Carbon Dioxide Emissions*, Washington, DC, September 2006, 5. Ex. E-188; see also Congressional Budget Office, *The Economics of Climate Change: A Primer*, Washington, DC, April 2003, 2. Ex. E-189 ("Inaction benefits people who are alive today while potentially harming future generations. Reducing emissions now may benefit future generations while imposing costs on the current population").

20. Congressional Budget Office, *Evaluating the Role of Prices*, 17.

21. Intergovernmental Panel on Climate Change, *Climate Change 2007: Synthesis Report Summary for Policymakers*, 2007, 5.

22. Office of Science and Technology Policy, Intergovernmental Panel on Climate Change Finalizes Report, February 2, 2007. Ex. E-190.

23. President George W. Bush, Remarks During a Meeting on Energy Security and Climate Change, September 28, 2007. Ex. E-191.

24. US Climate Change Science Program, *Scenarios of Greenhouse Gas Emissions and Atmospheric Concentrations*, Sub-report 2.1A of Synthesis and Assessment Product 2.1 by the U.S. Climate Change Science Program and the Subcommittee on Global Change Research, Washington, DC, 2007, 94–95. Ex. E-192.

25. US Climate Change Science Program, *Scenarios of Greenhouse Gas Emissions*, 3, 99.

26. US Climate Change Science Program, *Our Changing Planet: The U.S. Climate Change Science Program for Fiscal Year 2009*, A Report by the Climate Change Science Program and the Subcommittee on Global Change Research, Washington, DC, July 2008. Ex. E-193.

27. US Climate Change Science Program, *Our Changing Planet*, 25–26.

28. US Climate Change Science Program, 57–58.

29. US Climate Change Science Program, *Analyses of the Effects of Global Change on Human Health and Welfare and Human Systems*, A Report by the U.S. Climate Change Science Program and the Subcommittee on Global Change Research, Washington, DC, September 2008.

30. David Fahrenthold and Juliet Eilperin, "Warming Is Major Threat to Humans, EPA Warns," *Washington Post*, July 18, 2008. Ex. E-194.

31. Amanda Beck, "Carbon Cuts a Must to Halt Warming—US Scientists," *Reuters*, December 13, 2007. Ex. E-195.

32. James Hansen et al., "Target Atmospheric CO_2: Where Should Humanity Aim?," *Open Atmospheric Science Journal* 2 (2008): 217 (emphasis added).

33. The White House, Office of the Press Secretary, President Bush Discusses Global Climate Change, The Rose Garden, June 11, 2001. Ex. E-196.

34. James L. Connaughton, Chairman, White House Council on Environmental Quality, Statement for the Hearing Before the Committee on Commerce, Science, and Transportation, US Senate, July 11, 2002. Ex. E-197.

35. Jim Connaughton, Sharon Hays, and Harlan Watson, Press Briefing Via Conference Call by Senior Administration Officials on IPCC Report, November 16, 2007. Ex. E-198.

36. US Department of State, *U.S. Climate Action Report 2002*, 50–69.

37. US Department of State, 52–58, 64–69.

38. US Government Accountability Office, *Renewable Energy: Wind Power's Contribution to Electric Power Generation and Impact of Farms and Rural Communities*, GAO-04-756, Washington, DC, September 2004, 5. Ex. E-199.

39. US Government Accountability Office, *Renewable Energy: Wind Power's Contribution.*

40. Dr. Harlan Watson, "U.S. Climate Change Policy" (Presentation at US–Germany Bilateral Meeting, Berlin, Germany, August 12, 2005). Ex. E-200.

41. M. Mulligan, "Tackling Climate Change in the United States: The Potential Contribution from Wind Power," Preprint, Conference Paper, National Renewable Energy Laboratory, July 2006, 1, 5. Ex. E-201.

42. Congressional Budget Office, *Evaluating the Role of Prices*, 10.

43. Congressional Budget Office, 1.

44. US Government Accountability Office, *Advanced Energy Technologies: Key Challenges to Their Development and Deployment*, GAO-07–550T, Statement of Jim Wells, Testimony Before the Subcommittee on Energy and Water Development, Committee on Appropriations, House of Representatives, February 28, 2007, 10. Ex. E-202.

45. US Government Accountability Office, *Advanced Energy Technologies: Key Challenges*, 4.

46. US Government Accountability Office, *Advanced Energy Technologies: Budget Trends and Challenges for DOE's Energy R&D Program*, GAO-08-556T, Statement of Mark E. Gaffigan, Testimony Before the Subcommittee on Energy and Environment, Committee on Science and Technology, House of Representatives, March 5, 2008, 11. Ex. E-203.

47. US Government Accountability Office, *Department of Energy: Key Challenges Remain for Developing and Deploying Advanced Energy Technologies to Meet Future Needs*, GAO-07–106, December 2006. Ex. E-204.

48. US Government Accountability Office, *Department of Energy: Key Challenges*, 3 (describing how the low cost of fossil fuels does not include negative externalities, such as environmental harms).

49. Randy Randol, Facsimile to Phil Cooney, March 22, 2002. Ex. E-205.

50. Myron Ebell, Competitive Enterprise Institute, Email to Phil Cooney, June 3, 2002. Ex. E-206.

51. Ebell, email to Cooney.

52. Committee on Oversight and Government Reform, *Political Interference*, ii.

53. Committee on Oversight and Government Reform, 33.

54. Committee on Oversight and Government Reform, i.

55. Andrew C. Revkin, "Editor of Climate Reports Resigns," *New York Times*, June 11, 2005, https://www.nytimes.com/2005/06/11/us/national-briefing-washington-editor-of-climate-reports-resigns.html. Ex. E-207.

56. National Aeronautics and Space Administration, Office of Inspector General, *Investigative Summary Regarding Allegations that NASA Suppressed Climate Change Science and Denied Media Access to Dr. James E. Hansen, a NASA Scientist*, June 2008, 47. Ex. E-208.

57. The White House, President Bush Discusses Global Climate Change.

58. See National Research Council, *Implementing Climate and Global Change Research: A Review of the Final U.S. Climate Change Science Program Strategic Plan* (Washington, DC: National Academies Press, 2004), 1. Ex. E-209.

59. National Research Council, *Implementing Climate and Global Change Research*, 1.

60. President George W. Bush, Remarks Announcing the Clear Skies and Global Climate Change Initiatives in Silver Spring, Maryland, February 14, 2002. Ex. E-210.

61. James L. Connaughton and John H. Marburger, III, Open Letter on the President's Position on Climate Change, February 7, 2007. Ex. E-211.

62. Connaughton and Marburger, Open Letter on the President's Position.

63. Christine Todd Whitman, Administrator of the U.S. Environmental Protection Agency, Remarks at The Business Council, Washington, DC, February 22, 2001. Ex. E-212.

64. Statement of Administration Policy: S. 3036—Lieberman–Warner Climate Security Act, June 2, 2008. Ex. E-213.

65. Juliet Eilperin, "White House Tried to Silence EPA Proposal on Car Emissions," *Washington Post*, June 26, 2008. Ex. E-214.

66. Statement of Administration Policy: H.R. 6—Energy Policy Act of 2005, June 14, 2005. Ex. E-215.

67. Robert E. Fabricant, EPA General Counsel, Memorandum to Marianne L. Horinko, Acting Administrator, EPA's Authority to Impose Mandatory Controls to Address Global Climate Change under the Clean Air Act, August 28, 2003. Ex. E-216.

68. President George W. Bush, Remarks on Energy in Athens, Alabama, June 21, 2007. Ex. E-217.

69. Stephen L. Johnson, EPA Administrator, Letter to the President, January 31, 2008. Ex. E-218.

70. Ian Talley and Siobhan Hughes, "White House Blocks EPA Emissions Draft," *Wall Street Journal*, June 30, 2008, https://www.wsj.com/articles/SB121478564162114625. Ex. E-219.

71. US Environmental Protection Agency, Regulating Greenhouse Gas Emissions Under the Clean Air Act, Advanced Notice of Proposed Rulemaking, 73 Fed. Reg. 44354 (July 30, 2008). Ex. E-220; see also Fahrenthold and Eilperin, "Warming Is Major Threat" ("Last Friday, the EPA announced that it would solicit comments on the idea of regulating greenhouse gases under the federal Clean Air Act. But at the same time it released a lengthy preamble, with messages from EPA Administrator Stephen L. Johnson and four other Cabinet members, saying that this idea was ill-advised. Burnett said that this, too, was ordered by administration officials: 'We were told . . . that the [document] should not establish a path forward or a framework for regulation, but should emphasize the complexity of the challenge'").

72. President George W. Bush, Remarks on Energy and Climate Change, April 16, 2008. Ex. E-221.

73. Thomas Fuller and Peter Gelling, "Deadlock Stymies Global Climate Talks," *New York Times*, December 12, 2007, https://www.nytimes.com/2007/12/12/world/12climate.html. Ex. E-222.

74. Exec. Order No. 13211, Actions Concerning Regulations That Significantly Affect Energy Supply, Distribution, or Use, 66 Fed. Reg. 28355 (May 22, 2001). Ex. E-223.

75. Exec. Order No. 13211.

76. Mitchell E. Daniels Jr., Memorandum for Heads of Executive Departments and Agencies, and Independent Regulatory Agencies, Guidance for Implementing E.O. 13211, July 13, 2001, 2. Ex. E-224.

77. Exec. Order No. 13212, Actions to Expedite Energy-Related Projects, 66 Fed. Reg. 28357 (May 22, 2001). Ex. E-225.

78. President George W. Bush, Memorandum for the Vice President et al., National Energy Policy Development Group, January 29, 2001, 2. Ex. E-226.

79. National Energy Policy Development Group, *National Energy Policy*, Washington, DC, May 2001, xiii. Ex. E-227.

80. Michael Abramowitz and Steven Mufson, "Papers Detail Industry's Role in Cheney's Energy Report," *Washington Post*, July 18, 2007, http://www.washingtonpost.com/wp-dyn/content/article/2007/07/17/AR2007071701987.html. Ex. E-228.

81. National Energy Policy Development Group, *National Energy Policy*.

82. President George W. Bush, Remarks Announcing the Energy Plan in St. Paul, Minnesota, May 17, 2001. Ex. E-229.

83. The White House, Office of the Press Secretary, President Discusses Energy Policy, Franklin County Veterans Memorial, Columbus, Ohio, March 9, 2005. Ex. E-230.

84. "The Halliburton Loophole," *New York Times*, November 2, 2009, https://www.nytimes.com/2009/11/03/opinion/03tue3.html. Ex. E-231.

85. Statement of Administration Policy: H.R. 6.

86. Keith Hennessey and Bryan Hannegan, Memorandum to the President through Al Hubbard, Energy Policy—Interim Report, September 30, 2005. Ex. E-232.

87. Martin Kettle, "Cheney Tells US to Carry on Guzzling," *Guardian*, May 10, 2001, https://www.theguardian.com/world/2001/may/10/dickcheney.martinkettle. Ex. E-233.

CHAPTER 7

1. US Department of Defense, *2014 Climate Change Adaptation Roadmap*, 2014, Foreword. https://www.acq.osd.mil/EIE/Downloads/CCARprint_wForward_e.pdf.

2. National Intelligence Council, *Implications for US National Security of Anticipated Climate Change*, NIC WP 2016–01, September 21, 2016, 3, https://www.dni.gov/files/documents/Newsroom/Reports%20and%20Pubs/Implications_for_US_National_Security_of_Anticipated_Climate_Change.pdf.

3. *Massachusetts v. EPA*, 549 U.S. 497 (2007).

4. US Global Change Research Program, *Global Climate Change Impacts in the United States* (New York: Cambridge University Press, 2009), 100. Ex. E-234.

5. *Climate Change and National Security, Hearing before the Committee on Environment and Public Works*, 111th Congress (July 30, 2009). Ex. E-235.

6. *Climate Change Impacts on National Parks in Colorado, Hearing before the Subcommittee on National Parks of the Committee on Energy and Natural Resources*, 111th Congress (August 24, 2009). Ex. E-236.

7. US Geological Survey, "Science-Based Strategies for Sustaining Coral Ecosystems," Fact Sheet 2009–3089, September 2009, 1. Ex. E-237.

8. US Environmental Protection Agency, Endangerment and Cause or Contribute Findings for Greenhouse Gases Under Section 202(a) of the Clean Air Act, Final Rule, 74 Fed. Reg. 66496 (December 15, 2009). Ex. E-238.

9. US Environmental Protection Agency, Endangerment and Cause or Contribute Findings, 66498–99, 66517.

10. *Clean Energy Policies That Reduce our Dependence on Oil, Hearing before the Subcommittee on Energy and Environment of the Committee on Energy and Commerce*, 111th Congress 30–31 (April 28, 2010). Ex. E-239.

11. See, e.g., National Research Council, *Advancing the Science of Climate Change* (Washington, DC: National Academies Press, 2010), 227–28. Ex. E-240. This report addressed the issue of abrupt changes in the climate, or "surprises," when tipping points are crossed. Such abrupt changes could include rapid ice sheet disintegration (with a corresponding rise in sea levels), irreversible drying and desertification of the subtropics, the release of methane from permafrost or methane hydrates in the oceans, or rapid increases in temperatures.

12. National Research Council, *Climate Stabilization Targets: Emissions, Concentrations, and Impacts over Decades to Millennia* (Washington, DC: National Academies Press, 2011), 21, 52. Ex. E-241.

13. National Research Council, *Climate Stabilization Targets*, 9.

14. Congressional Research Service, *Climate Change: Conceptual Approaches and Policy Tools*, August 29, 2011, 5–9. Ex. E-242.

15. National Oceanic and Atmospheric Administration, Office of Coast Survey, "Historical Geographic Place Names Removed from NOAA Charts," Edition 3.0, updated August 4, 2014. https://historicalcharts.noaa.gov/pdfs/HistoricalPlacenames_Louisiana.pdf.

16. US Environmental Protection Agency, *Climate Change Indicators in the United States, 2012*, 2nd ed., Washington, DC, December 2012. Ex. E-243.

17. US Environmental Protection Agency, *Climate Change Indicators 2012*, 8.

18. US Environmental Protection Agency, *Climate Change Indicators in the United States, 2014*, 3rd ed., Washington, DC, May 2014. Ex. E-244.

19. US Environmental Protection Agency, *America's Children and the Environment, 3rd ed.*, Washington, DC, January 2013, 107. Ex. E-245.

20. Jerry M. Melillo, Terese Richmond, and Gary W. Yohe, eds., *Climate Change Impacts in the United States: The Third National Climate Assessment*, U.S. Global Change Research Program (Washington DC: US Government Printing Office, 2014), 1. Ex. E-246.

21. Melillo et al., *Climate Change Impacts: Third National Climate Assessment*, 15.

22. US Global Change Research Program, *The Impacts of Climate Change on Human Health in the United States: A Scientific Assessment*, Washington, DC, 2016, 2, 4. Ex. E-247.

23. US Global Change Research Program, *Impacts of Climate Change on Human Health*, 1.

24. US Environmental Protection Agency, *Inventory of U.S. Greenhouse Gas Emissions*.

25. US Energy Information Administration, *Monthly Energy Review March 2018*, 2018, Table 12.1.

26. Intergovernmental Panel on Climate Change, *Climate Change 2014 Synthesis Report: Summary for Policymakers*, 2014, 2, 8. https://www.ipcc.ch/pdf/assessment-report/ar5/syr/AR5_SYR_FINAL_SPM.pdf. Ex. E-248.

27. The White House, Office of the Press Secretary, Inaugural Address by President Barack Obama, January 21, 2013. Ex. E-249.

28. The White House, Office of the Press Secretary, Remarks by the President in the State of the Union Address, February 12, 2013. Ex. E-250.

29. Julie Hirschfeld Davis, Mark Landler, and Coral Davenport, "Obama on Climate Change: The Trends Are 'Terrifying,'" *New York Times*, September 8, 2016, https://www.nytimes.com/2016/09/08/us/politics/obama-climate-change.html. Ex. E-251.

30. Congressional Research Service, *Options for a Federal Renewable Electricity Standard*, November 12, 2010, 2. Ex. E-252.

31. Congressional Research Service, *Options for a Federal Renewable Electricity Standard*, 4.

32. US Department of Energy, *Strategic Plan*, May 2011, 10, 12–13. Ex. E-253.

33. National Renewable Energy Laboratory, *Renewable Electricity Futures Study Volume 1: Exploration of High-Penetration Renewable Electricity Futures* (Golden, CO: National Renewable Energy Laboratory, June 2012), xviii. Ex. E-254.

34. The White House, Inaugural Address by President Obama, 2013.

35. US Department of Energy, "Chapter I: Transforming the Nation's Electricity System: The Second Installment of the Quadrennial Energy Review," in *Transforming the Nation's Electricity System* (January 2017), 1–1. Ex. E-255.

36. Executive Office of the President, *The Cost of Delaying Action to Stem Climate Change*, July 2014, 2. Ex. E-256.

37. The Obama Administration's account of its climate record can be found at https://obamawhitehouse.archives.gov/the-record/climate.

38. US Environmental Protection Agency and National Highway Traffic Safety Administration, Light-Duty Vehicle Greenhouse Gas Emission Standards and Corporate Average Fuel Economy Standards; Final Rule, 75 Fed. Reg. 25325, 25369 (May 7, 2010).

39. US Environmental Protection Agency and National Highway Traffic Safety Administration, 2017 and Later Model Year Light-Duty Vehicle Greenhouse Gas Emissions and Corporate

Average Fuel Economy Standards: Final Rule, 77 Fed. Reg. 62624, 62627 (October 15, 2012).

40. See https://obamawhitehouse.archives.gov/the-record/climate.
41. See President Barack Obama, "Obama's Big Climate Action Plan Announcement," June 25, 2013, https://www.youtube.com/watch?v=TC17DJl6-Ck.
42. The White House, Office of the Press Secretary, Remarks by the President on Climate Change, Georgetown University, Washington, DC, June 25, 2013. Ex. E-257.
43. President Barack Obama, The President's Clean Power Plan. Ex. E-258.
44. The White House, Office of the Press Secretary, Fact Sheet: U.S. Reports Its 2025 Emissions Target to the UNFCCC, March 31, 2015. Ex. E-259.
45. Defendants' Answer to First Amended Complaint, January 13, 2017, ECF. No. 98 at para. 127.
46. Lisa Friedman and Brad Plumer, "E.P.A. Drafts Rule on Coal Plants to Replace Clean Power Plan," *New York Times*, July 5, 2018, https://www.nytimes.com/2018/07/05/climate/clean-power-plan-replacement.html. Ex. E-260.
47. United Nations, *Paris Agreement*, 2015, Article 2, https://unfccc.int/sites/default/files/english_paris_agreement.pdf. Ex. E-262.
48. United Nations Framework Convention on Climate Change, Report of the Conference of the Parties on its Fifteenth Session, Held in Copenhagen from 7 to 19 December 2009, https://unfccc.int/resource/docs/2009/cop15/eng/11a01.pdf. Ex. E-263.
49. Radoslav Dimitrov, "The Paris Agreement on Climate Change: Behind Closed Doors," *Global Environmental Politics* 16, no. 3 (2016): 1, 4 ("One important political development in Paris was the surge of countries who wanted to limit the global temperature rise to 1.5 degrees Celsius above pre-industrial levels. For the first time, now the majority (106 states, to be precise) demanded preventing a temperature rise of 1.5°C. Northern countries preferred 2 degrees instead. The U.S. was mildly opposed even to that, proposing that 2 degrees appear only in the preamble and not in the substantive sections of the treaty").
50. Dimitrov, "The Paris Agreement," 3.
51. The White House, *United States Mid-Century Strategy for Deep Decarbonization*, Washington, DC, November 2016.
52. The White House, Office of the Press Secretary, Fact Sheet: President Obama's 21st Century Clean Transportation System, February 4, 2016, https://obamawhitehouse.archives.gov/the-press-office/2016/02/04/fact-sheet-president-obamas-21st-century-clean-transportation-system.
53. Exec. Order No. 13514, Federal Leadership in Environmental, Energy, and Economic Performance, 74 Fed. Reg. 52117 (October 8, 2009). Ex. E-264.
54. Exec. Order No. 13653, Preparing the United States for the Impacts of Climate Change, 78 Fed. Reg. 66819 (November 6, 2013). Ex. E-265.

55. US Department of Energy, *Strategic Plan*, 2.
56. US Department of Energy, *Strategic Plan*, 2, 13.
57. Congressional Budget Office, *The Effects of Renewable or Clean Electricity Standards*, Washington, DC, July 2011, VII. Ex. E-266.
58. Executive Office of the President, *The Economics of Coal Leasing on Federal Lands: Ensuring a Fair Return to Taxpayers*, June 2016, 6. Ex. E-267.
59. US Department of Energy, "Chapter I: Transforming the Nation's Electricity System," 1–31.
60. Congressional Research Service, *U.S. Crude Oil Export Policy: Background and Considerations*, December 31, 2014, 2. Ex. E-268.
61. Congressional Research Service, *U.S. Crude Oil Export Policy*, 4.
62. The White House, Office of the Press Secretary, Remarks by the President in State of the Union Address, January 24, 2012. Ex. E-269.
63. The White House, Office of the Press Secretary, Remarks by the President on American-Made Energy, March 22, 2012. Ex. E-270.
64. The White House, Remarks by the President in the State of the Union 2013.

CHAPTER 8

1. Philip Bump, "Trump's Plan to Use a Cold War-Era Law to Bolster the Coal Industry, Explained," *Washington Post*, June 1, 2018, https://www.washingtonpost.com/news/politics/wp/2018/06/01/trumps-plan-to-use-a-cold-war-era-law-to-bolster-the-coal-industry-explained.
2. Kelly Levin and Dennis Tirpak, "This Month in Climate Science," World Resources Institute, Washington, DC, August 20, 2020, https://www.wri.org/blog/2020/08/month-climate-science-june-july-2020.
3. Under the Global Change Research Act of 1990, the Committee on Earth and Environmental Sciences of the Federal Coordinating Council on Science, Engineering, and Technology is to periodically prepare a scientific assessment, known as the NCA. Pub. L. No. 101–606, §106, 104 Stat. 3096, 3101 (1990) (codified at 15 U.S.C. §2936). The US Global Change Research Program coordinates and integrates the activities of thirteen participating federal departments and agencies that carry out research and support the nation's response to global change.
4. US Global Change Research Program, *Climate Science Special Report: Fourth National Climate Assessment, Volume I* (Washington, DC: 2017). Ex. E-271.
5. Doyle Rice, "Buried? Feds to Release Major Climate Report Day after Thanksgiving," *USA Today*, November 21, 2018, https://www.usatoday.com/story/news/nation/2018/11/21/climate-change-report-released-friday-after-thanksgiving/2080298002; Umair Irfan,

"Trump White House Issues Climate Change Report Undermining Its Own Policy," *Vox*, November 26, 2018, https://www.vox.com/2018/11/26/18112505/national-climate-assessment-2018-trump.

6. US Global Change Research Program, *Impacts, Risks, and Adaptation in the United States: Fourth National Climate Assessment, Volume II* (Washington, DC: 2018), https://nca2018.globalchange.gov.

7. Caitlin Oprysko, "'I Don't Believe It': Trump Dismisses Grim Government Report on Climate Change," *Politico*, November 26, 2018, https://www.politico.com/story/2018/11/26/trump-climate-change-report-1016494; Chris Cillizza, "Donald Trump Buried a Climate Change Report Because 'I Don't Believe It,'" *CNN*, November 27, 2018, https://www.cnn.com/2018/11/26/politics/donald-trump-climate-change/index.html.

8. M. D. Merrill, B. M. Sleeter, P. A. Freeman, J. Liu, P. D. Warwick, and B. C. Reed, *Federal Lands Greenhouse Gas Emissions and Sequestration in the United States: Estimates for 2005–14*, US Geological Survey Scientific Investigations Report 2018–5131, 2018, https://doi.org/10.3133/sir20185131.

9. Merrill et al., *Federal Lands Greenhouse Gas Emissions*, 1, Table 2.

10. Merrill et al., 8.

11. CBS News, "Trump Says Climate Change Not a 'Hoax' but Questions If It's 'Manmade,'" *CBS*, October 15, 2018, https://www.cbsnews.com/news/trump-says-climate-change-not-a-hoax-but-questions-if-its-manmade.

12. Washington Post Editorial Board, "Pruitt and Perry Continue to Play Down Climate Change," *Washington Post*, January 21, 2017, https://www.washingtonpost.com/opinions/pruitt-and-perry-continue-to-play-down-climate-change/2017/01/21/c891c61c-de97-11e6-ad42-f3375f271c9c_story.html.

13. Dino Grandoni, Brady Dennis, and Chris Mooney, "Scott Pruitt Asks Whether Global Warming is 'Necessarily is a Bad Thing,'" *Denver Post*, February 7, 2018, https://www.denverpost.com/2018/02/07/scott-pruitt-views-on-climate-change.

14. Juliet Eilperin, Josh Dawsey, and Brady Dennis, "White House Blocked Intelligence Agency's Written Testimony Calling Climate Change 'Possibly Catastrophic,'" *Washington Post*, June 8, 2019, https://www.washingtonpost.com/climate-environment/2019/06/08/white-house-blocked-intelligence-aides-written-testimony-saying-human-caused-climate-change-could-be-possibly-catastrophic.

15. Joel Clement, "I'm a Scientist. I'm Blowing the Whistle on the Trump Administration," *Washington Post*, July 19, 2017, https://www.washingtonpost.com/opinions/im-a-scientist-the-trumpadministration-reassigned-me-for-speaking-up-about-climatechange/2017/07/19/389b8dce-6b12-11e7-9c15-177740635e83_story.html; Juliet Eilperin, "Senate Democrats Call for an Investigation of Climate Scientist Whistleblower Complaint," *Washington Post*, July 24, 2017, https://www.washingtonpost.com/news/energy-environment/wp

/2017/07/24/senate-democrats-demand-ig-probe-of-interiors-reassigning-top-career-officials.

16. Lisa Friedman, "E.P.A. Cancels Talk on Climate Change by Agency Scientists," *New York Times*, October 22, 2017, https://www.nytimes.com/2017/10/22/climate/epa-scientists.html; Arianna Skibell, "Agency Keeps Scientists from Speaking at Watershed Conference," *Greenwire*, October 23, 2017, https://www.eenews.net/greenwire/2017/10/23/stories/1060064343.

17. Elliott Negin, "Energy Department Scientists Barred from Attending Nuclear Power Conference," *HuffPost*, August 1, 2017, https://www.huffpost.com/entry/energy-department-scientists-barred-from-attending_b_597f7e2ee4b0cb4fc1c73b8e; Brittany Patterson, "Govt. Scientist Blocked from Talking About Climate and Fire," *E&E News*, October 31, 2017, https://www.eenews.net/stories/1060065143; Sarah Kaplan, "Government Scientists Blocked from the Biggest Meeting in Their Field," *Washington Post*, December 22, 2017, https://www.washingtonpost.com/news/speaking-of-science/wp/2017/12/22/government-scientists-blocked-from-the-biggestmeeting-in-their-field; Dino Grandoni, "The Energy 202: Interior Agency Blocks Group of Archaeologists from Attending Scientific Conference," *Washington Post*, May 3, 2018, https://www.washingtonpost.com/news/powerpost/paloma/the-energy-202/2018/05/03/the-energy-202-interior-agency-blocks-group-of-archaeologists-from-attending-scientific-conference/5aea1d9230fb042db57972ac.

18. Alexander C. Kaufman, "Leaked Memo: EPA Shows Workers How to Downplay Climate Change," *HuffPost*, March 28, 2018, https://www.huffpost.com/entry/epa-climate-adaptation_n_5abbb5e3e4b04a59a31387d7.

19. Environmental Data & Governance Initiative, "Website Monitoring," https://envirodatagov.org/website-monitoring/; Environmental Data & Governance Initiative, "The New Digital Landscape: How the Trump Administration Has Undermined Federal Web Infrastructures for Climate Information," July 2019, https://envirodatagov.org/wp-content/uploads/2019/07/New_Digital_Landscape_EDGI_July_2019.pdf. See also "Climate Change," US Environmental Protection Agency, https://19january2017snapshot.epa.gov/climatechange_.html; Union of Concerned Scientists, "New UCS Report Tallies Attacks on Science in Trump Era Harming Public Health," January 28, 2019, https://www.ucsusa.org/about/news/ucs-report-tallies-attacks-science.

20. Nick Visser, "Interior Department Scrubs Climate Change From Agency Website. Again," *HuffPost*, June 13, 2017, https://www.huffpost.com/entry/interior-department-climate-change_n_593f8bcae4b0b13f2c6d8a9a.

21. Coral Davenport, "E.P.A. Faces Bigger Tasks, Smaller Budgets and Louder Critics," *New York Times*, March 18, 2016, https://www.nytimes.com/2016/03/19/us/politics/epa-faces-bigger-tasks-smaller-budgets-and-louder-critics.html.

22. Peter Stone, "'Swampy Symbiosis': Fossil Fuel Industry Has More Clout Than Ever Under Trump," *Guardian*, September 27, 2019, https://www.theguardian.com/environment/2019/sep/27/fossil-fuel-industry-clout-trump-era.

23. Environmental Integrity Project, "Who's Running Trump's EPA?," https://environmentalintegrity.org/trump-watch-epa/whos-running-trumps-epa.

24. Coral Davenport, Lisa Friedman, and Maggie Haberman, "E.P.A. Chief Scott Pruitt Resigns Under a Cloud of Ethics Scandals," *New York Times*, July 5, 2018, https://www.nytimes.com/2018/07/05/climate/scott-pruitt-epa-trump.html.

25. "Tracking How Many Key Positions Trump Has Filled So Far," *Washington Post*, https://www.washingtonpost.com/graphics/politics/trump-administration-appointee-tracker/database; Kathryn Dunn Tenpas, "Vacancies, Acting Officials and the Waning Role of the U.S. Senate," *The Brookings Institute*, September 24, 2020, https://www.brookings.edu/blog/fixgov/2020/09/24/vacancies-acting-officials-and-the-waning-role-of-the-u-s-senate.

26. Eric Levitz, "Trump Thwarts GOP Plot to Pretend His Climate Agenda Isn't Idiotic," *New York*, May 28, 2019, https://nymag.com/intelligencer/2019/05/trump-climate-denier-william-happer-co2-jews-science.html.

27. The 2018 to 2022 EPA strategic plan does not include goals and objectives related to climate change or discuss strategies for addressing the impacts of climate change effects. US Environmental Protection Agency, *Working Together: FY 2018–2022 U.S. EPA Strategic Plan*, EPA-190-R-18-003 (Washington, DC: February 2018), https://www.epa.gov/sites/production/files/2018-02/documents/fy-2018-2022-epa-strategic-plan.pdf. Moreover, neither the fiscal years 2018 to 2019 nor fiscal years 2020 to 2021 national program manager guidance for the EPA's Office of Land and Emergency Management mentions climate change among its goals and priorities. US Environmental Protection Agency, "Final FY 2018–2019 Office of Land and Emergency Management National Program Manager Guidance," 540B17001 (September 29, 2017), and "Final FY 2020–2021 Office of Land and Emergency Management National Program Manager Guidance," 500B19002 (June 7, 2019). The national program manager guidance communicates operational planning priorities, strategies, and key activities for advancing the agency's strategic plan.

28. FEMA, *2018–2022 Strategic Plan*, 2018, https://www.fema.gov/sites/default/files/2020-03/fema-strategic-plan_2018-2022.pdf.

29. US Government Accountability Office, *Climate Information: A National System Could Help Federal, State, Local, and Private Sector Decision Makers Use Climate Information*, GAO-16–37, Washington, DC, November 2015.

30. US Government Accountability Office, *High-Risk Series: Substantial Efforts Needed to Achieve Greater Progress on High-Risk Areas*, GAO-19–157SP, Washington, DC, March 2019.

31. Bruce Lieberman, "What Trump's Proposed NEPA Rollback Could Mean for the Climate," February 20, 2020, https://www.yaleclimateconnections.org/2020/02/what-trumps-proposed-nepa-rollback-could-mean-for-the-climate; Council on Environmental Quality, Update to the Regulations Implementing the Procedural Provisions of the National Environmental Policy Act, Final Rule, 85 Fed. Reg. 43304 (July 16, 2020), https://www.govinfo.gov/content/pkg/FR-2020-07-16/pdf/2020-15179.pdf.

32. See section 5 of Exec. Order No. 13783, Promoting Energy Independence and Economic Growth, 82 Fed. Reg. 16093 (March 31, 2017), https://www.govinfo.gov/content/pkg/FR-2017-03-31/pdf/2017-06576.pdf.

33. Andrew Wheeler, Memorandum to Assistant Administrators, Increasing Consistency and Transparency in Considering Benefits and Costs in the Rulemaking Process, May 13, 2019, https://www.epa.gov/environmental-economics/administrator-wheeler-memorandum-increasing-consistency-and-transparency.

34. US Government Accountability Office, *Social Cost of Carbon: Identifying a Federal Entity to Address the National Academies' Recommendations Could Strengthen Regulatory Analysis*, GAO-20-254, Washington, DC, June 2020.

35. "Carbon lock-in" refers to the dynamic whereby decisions relating to GHG-emitting technologies, infrastructure, and practices, as well as their supporting networks, render associated CO_2 emissions largely inevitable, making it more challenging, even impossible, to subsequently pursue paths toward low-carbon objectives. See, e.g., Peter Erickson, Sivan Kartha, Michael Lazarus, and Kevin Tempest, "Assessing Carbon Lock-In," *Environmental Research Letters* 10, no. 8 (2015). https://doi.org/10.1088/1748-9326/10/8/084023.

36. The following study shows how sensitive the attainment of emissions goals may be to near-term investment in new fossil fuel infrastructure: Christopher Smith et al., "Current Fossil Fuel Infrastructure Does Not Yet Commit Us to 1.5°C Warming," *Nature Communications* 10, no. 1 (2019), https://doi.org/10.1038/s41467-018-07999-w.

37. US Environmental Protection Agency, Oil and Natural Gas Sector: Emission Standards for New, Reconstructed, and Modified Sources: Stay of Certain Requirements: Proposed Rule, 82 Fed. Reg. 27645 (June 16, 2017), https://www.govinfo.gov/content/pkg/FR-2017-06-16/pdf/2017-12698.pdf (emphasis added.)

38. US Environmental Protection Agency, "EPA Issues Final Policy and Technical Amendments to the New Source Performance Standards for the Oil and Natural Gas Industry," https://www.epa.gov/controlling-air-pollution-oil-and-natural-gas-industry/epa-issues-final-policy-and-technical.

39. Dr. Jerome Paulson, Decl. in Support of Plaintiffs' Urgent Motion under Circuit Rule 27–3(b) for Preliminary Injunction, para. 23.

40. Paulson, Decl., para. 41.

41. Paulson, para. 34.

42. Paulson, paras. 27–30.

43. Dr. Lise Van Susteren, Decl. in Support of Plaintiffs' Urgent Motion under Circuit Rule 27–3(b) for Preliminary Injunction, paras. 13, 28.

44. Van Susteren, Decl., para. 12.

45. Aji P., Decl. in Support of Plaintiffs' Urgent Motion under Circuit Rule 27–3(b) for Preliminary Injunction, paras. 3–4, 8, 11; Levi D., Decl. in Support of Plaintiffs' Urgent

Motion under Circuit Rule 27–3(b) for Preliminary Injunction, paras. 7–9, 25; Journey Z., Decl. in Support of Plaintiffs' Urgent Motion under Circuit Rule 27–3(b) for Preliminary Injunction, paras. 25–26; Nicholas V., Decl. in Support of Plaintiffs' Urgent Motion under Circuit Rule 27–3(b) for Preliminary Injunction, para. 8.

46. See, e.g., Aji, Decl., para. 11; Levi, Decl., paras. 8, 24–25; Journey, Decl., para. 25; Nicholas, Decl., para. 8.

47. Van Susteren, Decl., para. 16.

48. Aji, Decl., para. 11.

49. Exec. Order No. 13783.

50. Marianne Lavelle, "Trump's Executive Order: More Fossil Fuels, Regardless of Climate Change," *Inside Climate News*, March 28, 2017, https://insideclimatenews.org/news/28032017/trump-executive-order-climate-change-paris-climate-agreement-clean-power-plan-pruitt.

51. Nicole Gentile and Kate Kelly, "The Trump Administration Is Stifling Renewable Energy on Public Lands and Waters," *Center for American Progress*, June 25, 2020, https://www.americanprogress.org/issues/green/reports/2020/06/25/486852/trump-administration-stifling-renewable-energy-public-lands-waters.

52. Office of Management and Budget, *A Budget for a Better America: Fiscal Year 2020 Budget of the U.S. Government* (Washington, DC: US Government Publishing Office, 2019), https://www.govinfo.gov/content/pkg/BUDGET-2020-BUD/pdf/BUDGET-2020-BUD.pdf.

53. Dan Gearino, "Trump's Budget Could Have Chilling Effect on U.S. Clean Energy Leadership," *Inside Climate News*, April 2, 2019, https://insideclimatenews.org/news/02042019/trump-budget-cuts-national-labs-clean-energy-leadership-solar-wind-electric-vehicles-climate-change.

54. Bianca Majumder, Sally Hardin, and Claire Moser, "The Impact of the Coronavirus on the Renewable Energy Industry," *Center for American Progress*, April 15, 2020, https://www.americanprogress.org/issues/green/news/2020/04/15/483219/impact-coronavirus-renewable-energy-industry.

55. Reuters, "Clean Energy Has Shed Nearly 600,000 U.S. Jobs Due to Pandemic: Report," *Reuters*, May 13, 2020, https://www.reuters.com/article/us-usa-jobs-clean-energy/clean-energy-has-shed-nearly-600000-us-jobs-due-to-pandemic-report-idUSKBN22P2TH.

56. Mark Jacobson, "Abstracts of 47 Peer-Reviewed Published Journal Articles From 13 Independent Research Groups With 91 Different Authors Supporting the Result That Energy for Electricity, Transportation, Building Heating/Cooling, and/or Industry Can Be Supplied Reliably with 100% or Near-100% Renewable Energy at Difference Locations Worldwide," December 29, 2019, http://web.stanford.edu/group/efmh/jacobson/Articles/I/CombiningRenew/100PercentPaperAbstracts.pdf.

57. Dr. Joseph Stiglitz, Decl. in Support of Plaintiffs' Urgent Motion under Circuit Rule 27–3(b) for Preliminary Injunction, para. 10 ("For decades, the U.S. government has had extensive knowledge that there were viable alternatives to a fossil fuel-based, national energy system, and with the appropriate allocation of further resources to research and development, it is likely that these alternatives would have been even more competitive than fossil fuels"); Peter Erickson, Decl. in Support of Plaintiffs' Urgent Motion under Circuit Rule 27–3(b) for Preliminary Injunction, para. 14 (Defendants have plans for "new offshore oil and gas drilling in virtually all (98%) of U.S. coastal waters during 2019–2024").

58. Erickson, Decl., para. 15.

59. Erickson, para. 18.

60. Vice Admiral Lee Gunn, Decl. in Support of Plaintiffs' Urgent Motion under Circuit Rule 27–3(b) for Preliminary Injunction, passim.

61. Dr. Steven Running, Decl. in Support of Plaintiffs' Urgent Motion under Circuit Rule 27–3(b) for Preliminary Injunction, para. 46.

62. The White House, "Energy & Environment," https://www.whitehouse.gov/issues/energy-environment. Ex. E-288.

63. Rhodium Group, "Preliminary US Emissions Estimates for 2018," January 8, 2019, https://rhg.com/research/preliminary-us-emissions-estimates-for-2018.

64. Rhodium Group, "Preliminary US Emissions Estimates."

65. IEA, *World Energy Investment 2018* (Paris, France: Organisation for Economic Co-operation and Development, 2018), https://www.iea.org/wei2018.

66. Exec. Order 13783.

67. US Department of the Interior, Bureau of Ocean Energy Management, *2019–2024 National Outer Continental Shelf Oil and Gas Leasing, Draft Proposed Program*, January 2018, https://www.boem.gov/sites/default/files/oil-and-gas-energy-program/Leasing/Five-Year-Program/2019-2024/DPP/NP-Draft-Proposed-Program-2019-2024.pdf.

68. US Department of the Interior, "Interior Announces Date for Largest Oil and Gas Lease Sale in U.S. History," February 16, 2018, https://www.doi.gov/pressreleases/interior-announces-date-largest-oil-and-gas-lease-sale-us-history.

69. Bureau of Land Management, "Interior's Bureau of Land Management Begins Planning Effort for NPR-A," November 20, 2018, https://www.blm.gov/press-release/interiors-bureau-land-management-begins-planning-effort-npr-a.

70. The White House, "President Donald J. Trump Is Unleashing American Energy Dominance," May 14, 2019, https://trumpwhitehouse.archives.gov/briefings-statements/president-donald-j-trump-unleashing-american-energy-dominance/.

71. US Environmental Protection Agency, "EPA Takes Another Step to Advance President Trump's America First Strategy, Proposes Repeal of 'Clean Power Plan,'" October 10, 2017, https://archive.epa.gov/epa/newsreleases/epa-takes-another-step-advance-president

-trumps-america-first-strategy-proposes-repeal.html; US Environmental Protection Agency, Repeal of the Clean Power Plan; Emission Guidelines for Greenhouse Gas Emissions From Existing Electric Utility Generating Units; Revisions to Emission Guidelines Implementing Regulations: Final Rule, 84 Fed. Reg. 32520 (July 8, 2019), https://www.federalregister.gov/documents/2019/07/08/2019-13507/repeal-of-the-clean-power-plan-emissionguidelines-for-greenhouse-gas-emissions-from-existing.

72. US Environmental Protection Agency, "The Safer Affordable Fuel Efficient (SAFE) Vehicles Proposed Rule for Model Years 2021–2026," https://www.epa.gov/regulations-emissions-vehicles-and-engines/safer-and-affordable-fuel-efficient-vehicles-proposed; Coral Davenport, "U.S. to Announce Rollback of Auto Pollution Rules, a Key Effort to Fight Climate Change," *New York Times*, March 30, 2020, https://www.nytimes.com/2020/03/30/climate/trump-fuel-economy.html; Jeff Alson, "The Pandemic Hasn't Stopped Trump's Rollback of Clean Car Standards," *The Hill*, March 31, 2020, https://thehill.com/opinion/energy-environment/490431-the-pandemic-hasnt-stopped-trumps-rollback-of-clean-car-standards; US Environmental Protection Agency, "Regulations for Greenhouse Gas Emissions from Passenger Cars and Trucks," https://www.epa.gov/regulations-emissions-vehicles-and-engines/regulations-greenhouse-gas-emissions-passenger-cars-and; National Highway Traffic Safety Administration, "Corporate Average Fuel Economy," https://www.nhtsa.gov/laws-regulations/corporate-average-fuel-economy.

73. Coral Davenport and Lisa Friedman, "E.P.A. Inspector General to Investigate Trump's Biggest Climate Rollback," *New York Times*, July 27, 2020, https://www.nytimes.com/2020/07/27/climate/trump-fuel-efficiency-rule.html.

74. US Department of the Interior, "Secretary Zinke Announces Plan For Unleashing America's Offshore Oil and Gas Potential," January 4, 2018, https://www.doi.gov/pressreleases/secretary-zinke-announces-plan-unleashing-americas-offshore-oil-and-gas-potential. Ex. E-273.

75. US Department of the Interior, "Secretary Bernhardt Signs Decision to Implement the Coastal Plain Oil and Gas Leasing Program in Alaska," August 17, 2020, https://www.doi.gov/pressreleases/secretary-bernhardt-signs-decision-implement-coastal-plain-oil-and-gas-leasing-program.

76. See, e.g., Exec. Order No. 13795, Implementing an America-First Offshore Energy Strategy, 82 Fed. Reg. 20815 (May 3, 2017). Ex. E-274 (revoking Exec. Order No. 13754, December 9, 2016); see also US Department of the Interior, Sec. Order No. 3350, America-First Offshore Energy Strategy, May 1, 2017. Ex. E-275; US Department of the Interior, "Secretary Zinke Announces Largest Oil & Gas Lease Sale in U.S. History," October 24, 2017, https://www.doi.gov/pressreleases/secretary-zinke-announces-largest-oil-gas-lease-sale-us-history. Ex. E-276.

77. See Brad Plumer, "Trump Orders a Lifeline for Struggling Coal and Nuclear Plants," *New York Times*, June 1, 2018, https://www.nytimes.com/2018/06/01/climate/trump-coal-nuclear-power.html. Ex. E-277; Jennifer A. Dlouhy, "Trump Prepares Lifeline for

Money-Losing Coal Plants," *Bloomberg*, May 31, 2018, https://www.bloomberg.com/news/articles/2018-06-01/trump-said-to-grant-lifeline-to-money-losing-coal-power-plants-jhv94ghl. Ex. E-278; Grid Memo, May 29, 2018, https://www.documentcloud.org/documents/4491203-Grid-Memo.html. Ex. E-279.

78. US Department of Energy, "DOE Announces Intent to Provide $122M to Establish Coal Products Innovation Centers," June 26, 2020, https://www.energy.gov/articles/doe-announces-intent-provide-122m-establish-coal-products-innovation-centers.

79. Bureau of Land Management, "BLM Offers Revision to Methane Waste Prevention Rule," February 12, 2018, https://www.blm.gov/press-release/blm-offers-revision-methane-waste-prevention-rule. Ex. E-280.

80. Coral Davenport, "Trump Eliminates Major Methane Rule, Even as Leaks Are Worsening," *New York Times*, August 20, 2020, https://www.nytimes.com/2020/08/13/climate/trump-methane.html; US Environmental Protection Agency, Oil and Natural Gas Sector: Emission Standards for New, Reconstructed, and Modified Sources Review: Final Rule, 85 Fed. Reg. 57018 (September 14, 2020), https://www.govinfo.gov/content/pkg/FR-2020-09-14/pdf/2020-18114.pdf.

81. Pipeline and Hazardous Materials Safety Administration, Pipeline Safety: Gas Pipeline Regulatory Reform, Notice of Proposed Rulemaking, 85 Fed. Reg. 35240 (June 9, 2020), https://www.govinfo.gov/content/pkg/FR-2020-06-09/pdf/2020-11843.pdf.

82. Ana Swanson and Brad Plumer, "Trump Slaps Steep Tariffs on Foreign Washing Machines and Solar Products," *New York Times*, January 22, 2018, https://www.nytimes.com/2018/01/22/business/trump-tariffs-washing-machines-solar-panels.html. Ex. E-281.

83. Jeff Tollefson, "U.S. Government Disbands Climate-Science Advisory Committee," *Scientific American*, April 21, 2017, https://www.scientificamerican.com/article/u-s-government-disbands-climate-science-advisory-committee. Ex. E-282.

84. Kevin Liptak and Jim Acosta, "Trump on Paris Accord: 'We're Getting Out,'" *CNN*, June 2, 2017, https://www.cnn.com/2017/06/01/politics/trump-paris-climate-decision/index.html. Ex. E-283.

85. The White House, *National Security Strategy of the United States of America*, December 2017. Ex. E-284.

86. Council on Environmental Quality, Withdrawal of Final Guidance for Federal Departments and Agencies on Consideration of Greenhouse Gas Emissions and the Effects of Climate Change in National Environmental Policy Act Reviews: Notice, 82 Fed. Reg. 16576 (April 5, 2017). Ex. E-285.

87. Exec. Order No. 13927, Accelerating the Nation's Economic Recovery From the COVID–19 Emergency by Expediting Infrastructure Investments and Other Activities, 85 Fed. Reg. 35165 (June 9, 2020), https://www.govinfo.gov/content/pkg/FR-2020-06-09/pdf/2020-12584.pdf.

88. Exec. Order No. 13766, Expediting Environmental Reviews and Approvals for High Priority Infrastructure Projects, 82 Fed. Reg. 8657 (January 30, 2017). Ex. E-286; Presidential Memorandum, Construction of the Keystone XL Pipeline, 82 Fed. Reg. 8663 (January 30, 2017). Ex. E-287.

89. US Department of Energy, Office of Fossil Energy, Extending Natural Gas Export Authorizations to Non-Free Trade Agreement Countries Through the Year 2050, 85 Fed. Reg. 52237 (August 25, 2020), https://www.energy.gov/sites/prod/files/2020/09/f78/2020-16836_FE_Policy%20Statement%20Year%202050.pdf.

90. Title II, Subtitle E, Pub. L. 104–121, 5 U.S.C. §§601 *et seq.*

91. 5 U.S.C. §801(a)(1)(A).

92. Congressional Research Service, *The Congressional Review Act (CRA): Frequently Asked Questions*, updated January 14, 2020, https://crsreports.congress.gov/product/pdf/R/R43992.

93. Julie Hirschfeld Davis, "Spending Plan Passed by Congress Is a Rebuke to Trump. Here's Why," *New York Times*, March 22, 2018, https://www.nytimes.com/2018/03/22/us/politics/trump-government-spending-bill.html.

94. Mark Hand, "Climate, Environmental Programs Left Mostly Untouched in Budget Deal," *Think Progress*, May 1, 2017, https://archive.thinkprogress.org/climate-environmental-programs-left-mostly-untouched-in-budget-deal-3742f7bad9c5/.

95. Science News, "Trump, Congress Approve Largest U.S. Research Spending Increase in a Decade," *Science*, March 23, 2018, http://www.sciencemag.org/news/2018/03/updated-us-spending-deal-contains-largest-research-spending-increase-decade.

96. Georgina Gustin, "Tax Overhaul Preserves Critical Credits for Wind, Solar and Electric Vehicles," *Inside Climate News*, December 22, 2017, https://insideclimatenews.org/news/18122017/tax-bill-vote-renewable-credits-solar-wind-clean-energy-jobs-evs-investment-anwr. This bill included a provision allowing drilling in part of the Arctic National Wildlife Refuge.

97. National Defense Authorization Act for Fiscal Year 2018, https://www.govinfo.gov/content/pkg/BILLS-115hr2810enr/pdf/BILLS-115hr2810enr.pdf.

98. H.Res. 109, Recognizing the Duty of the Federal Government to Create a Green New Deal, 116th Congress (2019); S.Res. 59, A Resolution Recognizing the Duty of the Federal Government to Create a Green New Deal, 116th Congress (2019).

99. H.Res. 6, Adopting the Rules of the House of Representatives for the One Hundred Sixteenth Congress, and for Other Purposes, Section 104(f), 116th Congress (2019).

100. House Select Committee on the Climate Crisis, "Solving the Climate Crisis: The Congressional Action Plan for a Clean Energy Economy and a Healthy, Resilient, and Just America," June 2020, https://climatecrisis.house.gov/sites/climatecrisis.house.gov/files/Climate%20Crisis%20Action%20Plan.pdf.

APPENDIX: THE PROCEDURAL HISTORY OF *JULIANA*

1. *In re United States*, 884 F.3d 830, 834 (9th Cir. 2018).
2. Declaration of Independence para. 2 (U.S. 1776).
3. James Madison, Address to the Agricultural Society of Albemarle, May 12, 1818.
4. *Juliana v. United States*, 339 F.Supp.3d 1062 (D. Or. 2018).
5. *Id.* at 1101.
6. *In re United States*, 139 S. Ct. 16 (Mem), 202 L. Ed. 2d 306 (2018).
7. *In re United States*, 139 S. Ct. 452 (Mem), 202 L. Ed. 2d 344 (2018).
8. *Juliana v. United States*, No. 18-80176 (9th Cir. Dec. 26, 2018) (order granting interlocutory appeal), https://perma.cc/CU97-WPYA.
9. *Id.* (Friedland, J., dissenting) (internal quotations omitted).
10. Those individuals and organizations filing *amicus* briefs in March 2019 included: Center for International Environmental Law and Environmental Law Alliance Worldwide-US; EarthRights International; Center for Biological Diversity; Defenders of Wildlife; the Union of Concerned Scientists; Food & Water Watch, Inc.; Friends of the Earth—US; Greenpeace, Inc.; International Lawyers for International Law; League of Women Voters U.S.; League of Women Voters of Oregon; Niskanen Center; Sierra Club; Sunrise Movement Education Fund; 82 Law Professors; 78 environmental history professors; businesses including Guayaki Yerba Mate, Royal Blue Organics, Organically Grown, Coconut Bliss, Hummingbird Wholesale, Aspen Skiing Company, and Protect Our Winters; members of the US Senate and House of Representatives (Sen. Ron Wyden of Oregon; Sen. Jeff Merkley of Oregon; Sen. Sheldon Whitehouse of Rhode Island; Rep. Debra Haaland of New Mexico; Reps. Peter DeFazio and Earl Blumenauer of Oregon; and Rep. Rashida Talib of Michigan); ministries (Eco-Justice Ministries; Interfaith Moral Action on Climate; General Synod of the United Church of Christ; Temple Beth Israel of Eugene, Oregon; National Advocacy Center of the Sisters of the Good Shepard; Leadership Council of the Sisters Servants of the Immaculate Heart of Mary of Monroe, Michigan; Sisters of Mercy of the Americas' Institute Leadership Team; GreenFaith; Leadership Team of the Sisters of Providence of Saint-Mary-of-the-Woods Indiana; Leadership Conference of Women Religious; Climate Change Task Force of the Sisters of Providence of Saint-Mary-of-the-Woods; Quaker Earthcare Witness; Colorado Interfaith Power and Light; and The Congregation of Our Lady of Charity of the Good Shepherd, U.S. Provinces); public health experts, public health organizations, and doctors (American Academy of Allergy, Asthma and Immunology; American Academy of Pediatrics; American Association of Community Psychiatrists; American Heart Association; American Lung Association; American Pediatric Society; American Thoracic Society; Infectious Diseases Society of America; International Society for Children's Health and the Environment; Medical Society Consortium on Climate and Health; National Association of County and City Health Officials; National

Environmental Health Association; National Medical Association; Society for Academic Emergency Medicine; and many preeminent experts in children's health).

11. *Juliana v. United States*, 947 F.3d 1159 (9th Cir. 2020).

12. Those individuals and organizations filing *amicus* briefs in March 2020 included: Senators Jeff Merkley of Oregon, Cory A. Booker of New Jersey, Tom Carper of Delaware, Edward J. Markey of Massachusetts, Brian Schatz of Hawaii, Chris Van Hollen of Maryland, Sheldon Whitehouse of Rhode Island, and Ron Wyden of Oregon; Representatives Jan Schakowsky of Illinois, Earl Blumenauer of Oregon, Yvette D. Clarke of New York, Judy Chu of California, Danny K. Davis of Illinois, Peter DeFazio of Oregon, Nanette Diaz Barragán of California, Adriano Espaillat of New York, Raúl Grijalva of Arizona, Debra Haaland of New Mexico, Jared Huffman of California, Sheila Jackson Lee of Texas, Barbara Lee of California, Jackie Speier of California, Rashida Tlaib of Michigan, and Debbie Wasserman Schultz of Florida; Amnesty International of the U.S.A.; Center for International Environmental Law; Environmental Law Alliance Worldwide—US; Global Justice Clinic at New York University School of Law; former Surgeons General Dr. Richard Carmona and Dr. David Satcher; League of Women Voters of the United States; League of Women Voters of Oregon; National Children's Campaign; Plaintiffs' expert witnesses in the litigation (Peter Erickson, Howard Frumkin, Ove Hoegh-Guldberg, Mark Jacobson, Susan Pacheco, Jerome Paulson, Eric Rignot, G. Philip Robertson, Steve Running, Catherine Smith, Gus Speth, Joseph Stiglitz, Harold Wanless, Kevin Trenberth, Lise Van Susteren, James Williams, and Andrea Wulf); Fred T. Korematsu Center for Law and Equality at Seattle University School of Law; Aoki Center for Critical Race and Nation Studies at UC Davis School of Law; Charles Hamilton Houston Institute for Race & Justice at Harvard University School of Law; Center on Race, Inequality, and the Law at New York University School of Law; Howard University Environmental Justice Center; children's rights advocates (Sacha M. Coupet, Nancy E. Dowd, Martha Albertson Fineman, Carmen G. Gonzalez, Shani M. King, Kalyani Robbins, Barbara Stark, Jonathan Todres, John Wall, Randee J. Waldman, Katherine Kaufka Walts, Tanya Washington, Barbara Bennett Woodhouse, the Center on Children and Families at the University of Florida Fredric G. Levin College of Law, the Child Rights Project of Emory Law School, the Juvenile Law Center); EarthRights International; Center for Biological Diversity; Defenders of Wildlife; Union of Concerned Scientists; Sierra Club; Food and Water Watch; Friends of the Earth-USA; law professors (Nadia B. Ahmad, Denise Antolini, Hope Babcock, Wil Burns, Alejandro E. Camacho, Erwin Chemerinsky, Kim Diana Connolly, Barbara Cosens, Erin Daly, Myanna Dellinger, Rachel E. Deming, Tim Duane, David L. Faigman, Richard M. Frank, Richard Hildreth, Keith H. Hirokawa, Helen H. Kang, Aliza Kaplan, Madeline June Kass, Robin Kundis Craig, Kenneth T. Kristl, Katrina Fischer Kuh, Douglas A. Kysar, James R. May, Patrick C. McGinley, Irma S. Russell, Erin Ryan, Mary Christina Wood); public health experts, public health organizations, and doctors (Samantha Ahdoot, Eric Chivian, Sir Andrew Haines, Kim Knowlton, Frederica Perera; American Academy of Allergy, Asthma and Immunology; American Academy of Pediatrics; American Association of Community

Psychiatrists; American Lung Association; American Pediatric Society; American Thoracic Society; International Society for Children's Health and the Environment; Medical Society Consortium on Climate and Health; National Association of County and City Health Officials; National Environmental Health Association; National Medical Association; Society for Academic Emergency Medicine; Edward Avol, Andra L. Blomkalns, Aparna Bole, David Brown, Jonathan M. Davis, Michael Donnenberg, Rani Gereige, Ronald Kleinman, Philip J. Landrigan, Jonathan Levy, Peter Orris, Jonathan Patz, James M. Perrin, Ali Raja, David Teitel, David Abramson, Laura Anderko, Rupa Basu, Susan Buchanan, Lori Byron, Richard Cash, Jiu-Chiuan Chen, Robin Cooper, Jeff Duchin, Christopher Golden, Jason Harris, Jeremy Hess, Matt Hollon, Julie R. Ingelfinger, Lindsay Jaacks, Richard J. Johnson, Bruce Lanphear, Regina LaRocque, Patrick Kinney, Yukari Manabe, Margaret M. McNamara, Mark Miller, Samuel Myers, Thomas B. Newman, Bart Ostro, Lori Peek, Bradley Peterson, Rebecca Philipsborn, Elizabeth G. Pinsky, David Pollack, James Recht, Irwin Redlener, Lynn Ringenberg, Jillan Sackett, Renee N. Salas, Mona Sarfaty, Shannon Scott-Vernaglia, Emily Senay, Perry Sheffield, Jonathan Slutzman, Caren Solomon, Gregory Wellenius, Mary L. Williams, Jonathan Winickoff).

References

5 U.S.C. §801(a)(1)(A).

15 U.S.C. §2902.

15 U.S.C. §2931(a)(1)–(2).

42 U.S.C. §§4342, 4343.

Abramowitz, Michael, and Steven Mufson. "Papers Detail Industry's Role in Cheney's Energy Report." *Washington Post*, July 18, 2007. http://www.washingtonpost.com/wp-dyn/content/article/2007/07/17/AR2007071701987.html.

Alson, Jeff. "The Pandemic Hasn't Stopped Trump's Rollback of Clean Car Standards." *The Hill*, March 31, 2020. https://thehill.com/opinion/energy-environment/490431-the-pandemic-hasnt-stopped-trumps-rollback-of-clean-car-standards.

Associated Press. "The Earth Summit; Excerpts From Speech By Bush on 'Action Plan.'" *New York Times*, June 13, 1992. https://www.nytimes.com/1992/06/13/world/the-earth-summit-excerpts-from-speech-by-bush-on-action-plan.html.

Ball, William L., Assistant to the President. Letter to Senator Gore. May 28, 1986.

Barney, Gerald O., ed. *Global 2000: The Report to the President: Entering the Twenty-First Century*. 1980, reissued in 1988. https://www.cartercenter.org/resources/pdfs/pdf-archive/global2000reporttothepresident--enteringthe21stcentury-01011991.pdf.

Bates, David, Assistant to the President and Secretary to the Cabinet. Draft Memorandum to Heads of Agencies. July 14, 1989.

Beck, Amanda. "Carbon Cuts a Must to Halt Warming—US Scientists." *Reuters*, December 13, 2007. https://www.reuters.com/article/idUSN13267425.

Bernthal, Frederick M. Letter to Professor Bert Bolin. July 5, 1990.

Bernthal, Frederick M. Memorandum to Richard T. McCormack Under Secretary-Designate for Economic Affairs, Attached Background Material. February 9, 1989.

Bernthal, Frederick M. Memorandum to the Secretary of the Department of State, Review of Key Foreign Policy Issues: The Environment. February 27, 1989.

Biello, David. "Where Did the Carter White House's Solar Panels Go?" *Scientific American*, August 6, 2010. https://www.scientificamerican.com/article/carter-white-house-solar-panel-array.

Briefing for Governor Sununu, Preparations for the First Framework Convention Negotiating Session, Washington, DC, February 4, 1991. November 7, 1990.

Bromley, D. Allen. Memorandum to Governor Sununu, The NRC Evans Report. April 8, 1991.

Bromley, D. Allen. Memorandum to John H. Sununu, Possible Group from Whom to Obtain Reaction to the NAS/Evans Report. April 18, 1991.

Bromley, D. Allen. Memorandum to John H. Sununu, Sea Level Change and Environmental Matters. June 14, 1990.

Bromley, D. Allan, and Ede Holiday. Memorandum for Governor Sununu, Climate Change Framework Convention Meeting at 9:30 Wednesday morning. January 23, 1991.

Bromley, Allan, and Stephen Dazansky. Memorandum for Fred Bernthal et al., Meeting of Global Change Strategy Task Force. May 21, 1990.

Bump, Philip. "Trump's Plan to Use a Cold War-Era Law to Bolster the Coal Industry, Explained." *Washington Post*, June 1, 2018. https://www.washingtonpost.com/news/politics/wp/2018/06/01/trumps-plan-to-use-a-cold-war-era-law-to-bolster-the-coal-industry-explained.

Bureau of Land Management. "BLM Offers Revision to Methane Waste Prevention Rule." February 12, 2018. https://www.blm.gov/press-release/blm-offers-revision-methane-waste-prevention-rule.

Bureau of Land Management. "Interior's Bureau of Land Management Begins Planning Effort for NPR-A." November 20, 2018. https://www.blm.gov/press-release/interiors-bureau-land-management-begins-planning-effort-npr-a.

Bush, George H. W. Presidential Address: Intergovernmental Panel on Climate Change, Georgetown University. February 5, 1990.

Bush, George H. W. Remarks on Signing the Energy Policy Act of 1992 in Maurice, Louisiana, October 24, 1992. https://www.presidency.ucsb.edu/node/267260.

Bush, George H. W. Remarks on Signing the Natural Gas Wellhead Decontrol Act of 1989. July 26, 1989. https://www.presidency.ucsb.edu/node/263165.

Bush, George W. Memorandum for the Vice President et al., National Energy Policy Development Group. January 29, 2001.

Bush, George W. Remarks Announcing the Clear Skies and Global Climate Change Initiatives in Silver Spring, Maryland. February 14, 2002. https://www.presidency.ucsb.edu/node/216061.

Bush, George W. Remarks Announcing the Energy Plan in St. Paul, Minnesota. May 17, 2001. https://www.presidency.ucsb.edu/node/213235.

Bush, George W. Remarks During a Meeting on Energy Security and Climate Change. September 28, 2007. https://www.presidency.ucsb.edu/node/276740.

Bush, George W. Remarks on Energy and Climate Change. April 16, 2008. https://www.presidency.ucsb.edu/node/206635.

Bush, George W. Remarks on Energy in Athens, Alabama. June 21, 2007. https://www.presidency.ucsb.edu/node/275685.

Cannon, Jonathan Z. Memorandum to Carol M. Browner, EPA's Authority to Regulate Pollutants Emitted by Electric Power Generation Sources. April 10, 1998. https://www.eenews.net/features/documents/2008/08/04/document_gw_05.pdf.

Carbon Dioxide and Climate: The Greenhouse Effect, Hearing before the Subcommittee on Natural Resources, Agricultural Research and Environment and the Subcommittee on Investigations and Oversight of the Committee on Science and Technology. 97th Congress. March 25, 1982.

Carr, Donald A., Acting Assistant Attorney General. Memorandum to Kenneth Yale, CEQ Global Climate Change Directive. June 23, 1989.

Carter, Jimmy. Address to the Nation on Energy and National Goals: "The Malaise Speech." July 15, 1979. https://www.presidency.ucsb.edu/node/249458.

Carter, Jimmy. Golden, Colorado Remarks at the Solar Energy Research Institute on South Table Mountain. May 3, 1978. https://www.presidency.ucsb.edu/node/245605.

Carter, Jimmy. National Energy Program Fact Sheet on the President's Program. April 20, 1977. https://www.presidency.ucsb.edu/node/243451.

Carter, Jimmy. Outer Continental Shelf Lands Act Amendments of 1978 Statement on Signing S.9 Into Law. September 18, 1978. https://www.presidency.ucsb.edu/node/243108.

Carter, Jimmy. Second Environmental Decade Remarks at the 10th Anniversary Observance of the National Environmental Policy Act, Earth Day, and Several Federal Agencies. February 29, 1980. https://www.presidency.ucsb.edu/node/250475.

Carter, Jimmy. Solar Energy Message to the Congress. June 20, 1979. https://www.presidency.ucsb.edu/node/250298.

Carter, Jimmy. The State of the Union Annual Message to the Congress. January 16, 1981. https://www.presidency.ucsb.edu/node/250760.

CBS News. "Trump Says Climate Change Not a 'Hoax' but Questions If It's 'Manmade.'" *CBS*, October 15, 2018. https://www.cbsnews.com/news/trump-says-climate-change-not-a-hoax-but-questions-if-its-manmade.

Cillizza, Chris. "Donald Trump Buried a Climate Change Report Because 'I Don't Believe It.'" *CNN*, November. 27, 2018. https://www.cnn.com/2018/11/26/politics/donald-trump-climate-change/index.html.

Clean Energy Policies That Reduce our Dependence on Oil, Hearing before the Subcommittee on Energy and Environment of the Committee on Energy and Commerce. 111th Congress. April 28, 2010. https://www.govinfo.gov/content/pkg/CHRG-111hhrg76568/pdf/CHRG-111hhrg76568.pdf.

Clement, Joel. "I'm a Scientist. I'm Blowing the Whistle on the Trump Administration." *Washington Post*, July 19, 2017. https://www.washingtonpost.com/opinions/im-a-scientist-the-trumpadministration-reassigned-me-for-speaking-up-about-climatechange/2017/07/19/389b8dce-6b12-11e7-9c15-177740635e83_story.html.

Climate Change and National Security, Hearing before the Committee on Environment and Public Works. 111th Congress. July 30, 2009. https://www.govinfo.gov/content/pkg/CHRG-111shrg99882/pdf/CHRG-111shrg99882.pdf.

Climate Change Impacts on National Parks in Colorado, Hearing before the Subcommittee on National Parks of the Committee on Energy and Natural Resources. 111th Congress. August 24, 2009. https://www.govinfo.gov/content/pkg/CHRG-111shrg52524/pdf/CHRG-111shrg52524.pdf.

Climate Change: Understanding the Degree of the Problem, Hearing before the Committee on Government Reform. 109th Congress. July 20, 2006. https://www.govinfo.gov/content/pkg/CHRG-109hhrg29932/pdf/CHRG-109hhrg29932.pdf.

Climate Research Board, National Research Council. *Carbon Dioxide and Climate: A Scientific Assessment.* Washington, DC: National Academy of Sciences, 1979.

Climate Surprises, Hearing before the Subcommittee on Science, Technology, and Space of the Committee on Commerce, Science, and Transportation. 101st Congress. May 8, 1989.

Clinton, William J. Remarks at the White House Conference on Climate Change. October 19, 1993. https://www.presidency.ucsb.edu/node/218750.

Clinton, William J. Remarks on Earth Day. April 21, 1993. https://www.presidency.ucsb.edu/node/219947.

Clinton, William J. Remarks on the 25th Observance of Earth Day in Havre de Grace, Maryland. April 21, 1995. https://www.presidency.ucsb.edu/node/220846.

Clinton, William J., and Al Gore, Jr. *The Climate Change Action Plan.* Washington DC: Executive Office of the President, 1993.

Coggins, George Cameron, and Doris K. Nagel. "'Nothing Beside Remains': The Legal Legacy of James G. Watt's Tenure as a Secretary of the Interior on Federal Land and Law Policy." *Boston College Environmental Affairs Law Review.* 17, no. 3 (1990): 473–550.

Committee on Oversight and Government Reform, US House of Representatives. *Political Interference with Climate Change Science Under the Bush Administration.* December 2007.

Congressional Budget Office. *Evaluating the Role of Prices and R&D in Reducing Carbon Dioxide Emissions.* Washington, DC, September 2006.

Congressional Budget Office. *The Economics of Climate Change: A Primer.* Washington, DC, April 2003. https://www.cbo.gov/sites/default/files/108th-congress-2003-2004/reports/04-25-climatechange.pdf.

Congressional Budget Office. *The Effects of Renewable or Clean Electricity Standards.* Washington, DC, July 2011. https://www.cbo.gov/sites/default/files/cbofiles/ftpdocs/121xx/doc12166/07-26-energy.pdf.

Congressional Research Service. *Climate Change: Conceptual Approaches and Policy Tools.* August 29, 2011. https://crsreports.congress.gov/product/pdf/R/R41973.

Congressional Research Service. *Options for a Federal Renewable Electricity Standard.* November 12, 2010. https://crsreports.congress.gov/product/pdf/R/R41493.

Congressional Research Service. *The Congressional Review Act (CRA): Frequently Asked Questions.* Updated January 14, 2020. https://crsreports.congress.gov/product/pdf/R/R43992.

Congressional Research Service. *U.S. Crude Oil Export Policy: Background and Considerations.* December 31, 2014. https://crsreports.congress.gov/product/pdf/R/R43442.

Congressional Research Service. *U.S. Greenhouse Gas Emissions: Recent Trends and Factors.* November 24, 2014. https://crsreports.congress.gov/product/pdf/R/R43795.

Connaughton, James L., Chairman, White House Council on Environmental Quality. Statement for the Hearing Before the Committee on Commerce, Science, and Transportation, US Senate. July 11, 2002.

Connaughton, James L., and John H. Marburger, III. Open Letter on the President's Position on Climate Change. February 7, 2007. https://georgewbush-whitehouse.archives.gov/news/releases/2007/02/20070207-5.html.

Connaughton, Jim, Sharon Hays, and Harlan Watson. Press Briefing Via Conference Call by Senior Administration Officials on IPCC Report. November 16, 2007. https://www.presidency.ucsb.edu/node/276961.

Council on Environmental Quality. *Environmental Quality 1977: The Eighth Annual Report of the Council on Environmental Quality.* Washington, DC, December 1977.

Council on Environmental Quality. *Environmental Quality 1980: The Eleventh Annual Report of the Council on Environmental Quality.* Washington, DC, December 1980.

Council on Environmental Quality. *Environmental Quality 1981: 12th Annual Report of the Council on Environmental Quality.* Washington, DC, 1982.

Council on Environmental Quality. *Environmental Quality 1982: 13th Annual Report of the Council on Environmental Quality.* Washington, DC.

Council on Environmental Quality. *Environmental Quality 1983: 14th Annual Report of the Council on Environmental Quality.* Washington, DC.

Council on Environmental Quality. *Environmental Quality: 15th Annual Report of the Council on Environmental Quality.* Washington, DC.

Council on Environmental Quality. *Environmental Quality: Along the American River: The 1996 Report of the Council on Environmental Quality.* Washington, DC.

Council on Environmental Quality. *Environmental Quality: The 1997 Report of the Council on Environmental Quality.* Washington, DC.

Council on Environmental Quality. *Environmental Quality: The First Annual Report of the Council on Environmental Quality together with the President's Message to Congress.* Washington, DC, August 1970.

Council on Environmental Quality. *Environmental Quality: The Ninth Annual Report of the Council on Environmental Quality.* Washington, DC, December 1978.

Council on Environmental Quality. *Environmental Quality: The Tenth Annual Report of the Council on Environmental Quality.* Washington, DC, 1979.

Council on Environmental Quality. *Environmental Quality: The Twentieth Annual Report of the Council on Environmental Quality together with The President's Message to Congress*. Washington, DC, 1990.

Council on Environmental Quality. *Environmental Quality: The Twenty-Third Annual Report of the Council on Environmental Quality together with The President's Message to Congress*. Washington, DC, January 1993.

Council on Environmental Quality. *Global Energy Futures and the Carbon Dioxide Problem*. Washington, DC, January 1981.

Council on Environmental Quality. *Solar Energy: Progress and Promise*. Washington, DC, April 1978. https://files.eric.ed.gov/fulltext/ED164264.pdf.

Council on Environmental Quality. *The Good News About Energy*. Washington, DC, 1979.

Council on Environmental Quality. Update to the Regulations Implementing the Procedural Provisions of the National Environmental Policy Act, Final Rule. 85 Fed. Reg. 43304, July 16, 2020. https://www.govinfo.gov/content/pkg/FR-2020-07-16/pdf/2020-15179.pdf.

Council on Environmental Quality. Withdrawal of Final Guidance for Federal Departments and Agencies on Consideration of Greenhouse Gas Emissions and the Effects of Climate Change in National Environmental Policy Act Reviews: Notice. 82 Fed. Reg. 16576, April 5, 2017. https://www.gpo.gov/fdsys/pkg/FR-2017-04-05/pdf/2017-06770.pdf.

Council on Environmental Quality and US Department of State. *Global Future: Time To Act*. Washington, DC, January 1981.

Curlee, Dotty, Department of Energy Action Officer. Report of Hearing, Joint Hearing on Carbon Dioxide and the Greenhouse Effect. March 25, 1982.

Cushman, John H., Jr. "U.S. Signs a Pact to Reduce Gases Tied to Warming." *New York Times*, November 13, 1998. https://www.nytimes.com/1998/11/13/world/us-signs-a-pact-to-reduce-gases-tied-to-warming.html.

Daniels, Mitchell E., Jr. Memorandum for Heads of Executive Departments and Agencies, and Independent Regulatory Agencies, Guidance for Implementing E.O. 13211. July 13, 2001.

Danzansky, Steve. Memorandum for Chris Dawson et al., Letter on IPCC Conclusions Paper. August 7, 1990.

Davenport, Carol. "E.P.A. Faces Bigger Tasks, Smaller Budgets and Louder Critics." *New York Times*, March 18, 2016. https://www.nytimes.com/2016/03/19/us/politics/epa-faces-bigger-tasks-smaller-budgets-and-louder-critics.html.

Davenport, Coral. "Trump Eliminates Major Methane Rule, Even as Leaks Are Worsening." *New York Times*, August 20, 2020. https://www.nytimes.com/2020/08/13/climate/trump-methane.html.

Davenport, Coral. "U.S. to Announce Rollback of Auto Pollution Rules, a Key Effort to Fight Climate Change." *New York Times*, March 30, 2020. https://www.nytimes.com/2020/03/30/climate/trump-fuel-economy.html.

Davenport, Coral, and Lisa Friedman. "E.P.A. Inspector General to Investigate Trump's Biggest Climate Rollback." *New York Times*, July 27, 2020. https://www.nytimes.com/2020/07/27/climate/trump-fuel-efficiency-rule.html.

Davenport, Carol, Lisa Friedman, and Maggie Haberman. "E.P.A. Chief Scott Pruitt Resigns Under a Cloud of Ethics Scandals." *New York Times*, July 5, 2018. https://www.nytimes.com/2018/07/05/climate/scott-pruitt-epa-trump.html.

David, Edward E., Jr. Memorandum for Peter Flanigan. October 20, 1970.

Davis, Julie Hirschfeld. "Spending Plan Passed by Congress Is a Rebuke to Trump. Here's Why." *New York Times*, March 22, 2018. https://www.nytimes.com/2018/03/22/us/politics/trump-government-spending-bill.html.

Davis, Julie Hirschfeld, Mark Landler, and Coral Davenport. "Obama on Climate Change: The Trends Are 'Terrifying.'" *New York Times*, September 8, 2016. https://www.nytimes.com/2016/09/08/us/politics/obama-climate-change.html.

Deep Water Royalty Relief Act, Pub. L. 104–58. November 28, 1995.

Defendants' Answer to First Amended Complaint. January 13, 2017. ECF. No. 98.

Department of Energy Organization Act of 1977, Pub. L. 95–91, 91 Stat. 565. enacted August 4, 1977.

Dimitrov, Radoslav. "The Paris Agreement on Climate Change: Behind Closed Doors." *Global Environmental Politics* 16, no. 3 (2016): 1–11.

Dlouhy, Jennifer A. "Trump Prepares Lifeline for Money-Losing Coal Plants." *Bloomberg*, May 31, 2018. https://www.bloomberg.com/news/articles/2018-06-01/trump-said-to-grant-lifeline-to-money-losing-coal-power-plants-jhv94ghl.

Domestic Policy Council. Domestic Gas and Oil Incentives. 1994.

DuBridge, Lee. Memorandum to John Ehrlichman, Recommendation to Pursue Nonconventional Vehicle Development. December 19, 1969.

Ebell, Myron, Competitive Enterprise Institute. Email to Phil Cooney. June 3, 2002.

Eilperin, Juliet. "Senate Democrats Call for an Investigation of Climate Scientist Whistleblower Complaint." *Washington Post*, July 24, 2017. https://www.washingtonpost.com/news/energy-environment/wp/2017/07/24/senate-democrats-demand-ig-probe-of-interiors-reassigning-top-career-officials.

Eilperin, Juliet. "White House Tried to Silence EPA Proposal on Car Emissions." *Washington Post*, June 26, 2008. https://www.washingtonpost.com/wp-dyn/content/article/2008/06/25/AR2008062502713.html.

Eilperin, Juliet and Darla Cameron. "How Trump Is Rolling Back Obama's Legacy." *Washington Post*, updated January 20, 2018. https://www.washingtonpost.com/graphics/politics/trump-rolling-back-obama-rules/.

Eilperin, Juliet, Josh Dawsey, and Brady Dennis. "White House Blocked Intelligence Agency's Written Testimony Calling Climate Change 'Possibly Catastrophic.'" *Washington Post,* June 8, 2019. https://www.washingtonpost.com/climate-environment/2019/06/08/white-house-blocked-intelligence-aides-written-testimony-saying-human-caused-climate-change-could-be-possibly-catastrophic.

Energy Policy Act of 1992, Pub. L. 102–486. October 24, 1992.

Energy Policy and Conservation Act of 1975, Pub. L. 94–163, 89 Stat. 871, Sec. 502(a)(1), enacted December 22, 1975.

Energy Policy Implications of Global Warming, Hearings before the Subcommittee on Energy and Power of the Committee on Energy and Commerce. 100th Congress. July 7 and September 22, 1988.

Environmental Data & Governance Initiative. "The New Digital Landscape: How the Trump Administration Has Undermined Federal Web Infrastructures for Climate Information." July 2019. https://envirodatagov.org/wp-content/uploads/2019/07/New_Digital_Landscape_EDGI_July_2019.pdf.

Environmental Data & Governance Initiative. "Website Monitoring." https://envirodatagov.org/website-monitoring.

Environmental Integrity Project. "Trump Watch: EPA." https://environmentalintegrity.org/trump-watch-epa.

Environmental Integrity Project. "Who's Running Trump's EPA?" https://environmentalintegrity.org/trump-watch-epa/whos-running-trumps-epa.

Erickson, Peter, Sivan Kartha, Michael Lazarus, and Kevin Tempest. "Assessing Carbon Lock-In." *Environmental Research Letters* 10, no. 8 (2015). https://doi.org/10.1088/1748-9326/10/8/084023.

Exec. Order No. 13045, Protection of Children from Environmental Health Risks and Safety Risks. 62 Fed. Reg. 19885. April 23, 1997. https://www.govinfo.gov/content/pkg/FR-1997-04-23/pdf/97-10695.pdf.

Exec. Order No. 13211, Actions Concerning Regulations that Significantly Affect Energy Supply, Distribution, or Use. 66 Fed. Reg. 28355. May 22, 2001. https://www.govinfo.gov/content/pkg/FR-2001-05-22/pdf/01-13116.pdf.

Exec. Order No. 13212, Actions to Expedite Energy-Related Projects. 66 Fed. Reg. 28357. May 22, 2001. https://www.govinfo.gov/content/pkg/FR-2001-05-22/pdf/01-13117.pdf.

Exec. Order No. 13514, Federal Leadership in Environmental, Energy, and Economic Performance. 74 Fed. Reg. 52117. October 8, 2009. https://www.govinfo.gov/content/pkg/FR-2009-10-08/pdf/E9-24518.pdf.

Exec. Order No. 13653, Preparing the United States for the Impacts of Climate Change. 78 Fed. Reg. 66819. November 6, 2013. https://www.govinfo.gov/content/pkg/FR-2013-11-06/pdf/2013-26785.pdf.

Exec. Order No. 13766, Expediting Environmental Reviews and Approvals for High Priority Infrastructure Projects. 82 Fed. Reg. 8657. January 30, 2017. https://www.govinfo.gov/content/pkg/FR-2017-01-30/pdf/2017-02029.pdf.

Exec. Order No. 13783, Promoting Energy Independence and Economic Growth. 82 Fed. Reg. 16093. March 31, 2017. https://www.govinfo.gov/content/pkg/FR-2017-03-31/pdf/2017-06576.pdf.

Exec. Order No. 13795, Implementing an America-First Offshore Energy Strategy. 82 Fed. Reg. 20815. May 3, 2017. https://www.govinfo.gov/content/pkg/FR-2017-05-03/pdf/2017-09087.pdf.

Exec. Order No. 13927, Accelerating the Nation's Economic Recovery From the COVID–19 Emergency by Expediting Infrastructure Investments and Other Activities. 85 Fed. Reg. 35165. June 9, 2020. https://www.govinfo.gov/content/pkg/FR-2020-06-09/pdf/2020-12584.pdf.

Executive Office of the President. *The Cost of Delaying Action to Stem Climate Change.* July 2014. https://obamawhitehouse.archives.gov/sites/default/files/docs/the_cost_of_delaying_action_to_stem_climate_change.pdf.

Executive Office of the President. *The Economics of Coal Leasing on Federal Lands: Ensuring a Fair Return to Taxpayers.* June 2016. https://obamawhitehouse.archives.gov/sites/default/files/page/files/20160622_cea_coal_leasing.pdf.

Fabricant, Robert E., EPA General Counsel. Memorandum to Marianne L. Horniko, Acting Administrator, EPA's Authority to Impose Mandatory Controls to Address Global Climate Change under the Clean Air Act. August 28, 2003.

Fahrenthold, David, and Juliet Eilperin. "Warming Is Major Threat to Humans, EPA Warns." *Washington Post,* July 18, 2008. http://www.washingtonpost.com/wp-dyn/content/article/2008/07/17/AR2008071701557.html.

FEMA. *2018–2022 Strategic Plan.* 2018. https://www.fema.gov/sites/default/files/2020-03/fema-strategic-plan_2018-2022.pdf.

Flowers, L. T., and P. J. Dougherty. "Wind Powering America: Goals, Approach, Perspectives, and Prospects." Preprint, National Renewable Energy Laboratory, March 2002. https://www.nrel.gov/docs/fy02osti/32097.pdf.

Fox, J. Edward. Letter to Senator Chafee. July 29, 1988.

Friedman, Lisa. "E.P.A. Cancels Talk on Climate Change by Agency Scientists." *New York Times,* October 22, 2017. https://www.nytimes.com/2017/10/22/climate/epa-scientists.html.

Friedman, Lisa, and Brad Plumer. "E.P.A. Drafts Rule on Coal Plants to Replace Clean Power Plan." *New York Times,* July 5, 2018. https://www.nytimes.com/2018/07/05/climate/clean-power-plan-replacement.html.

Fuller, Thomas, and Peter Gelling. "Deadlock Stymies Global Climate Talks." *New York Times,* December 12, 2007. https://www.nytimes.com/2007/12/12/world/12climate.html.

Gächter, Rolando A. *Federal Offshore Statistics: 1995: Leasing, Exploration, Production, and Revenue as of December 31, 1995*. Herndon, VA: US Department of Interior, 1997.

Gearino, Dan. "Trump's Budget Could Have Chilling Effect on U.S. Clean Energy Leadership." *Inside Climate News*, April 2, 2019. https://insideclimatenews.org/news/02042019/trump-budget-cuts-national-labs-clean-energy-leadership-solar-wind-electric-vehicles-climate-change.

Gentile, Nicole, and Kate Kelly. "The Trump Administration Is Stifling Renewable Energy on Public Lands and Waters." *Center for American Progress*, June 25, 2020. https://www.americanprogress.org/issues/green/reports/2020/06/25/486852/trump-administration-stifling-renewable-energy-public-lands-waters.

Gilfillan, D., G. Marland, T. Boden, and R. Andres. *Global, Regional, and National Fossil-Fuel CO_2 Emissions*. Carbon Dioxide Information Analysis Center, Appalachian State University. https://energy.appstate.edu/research/work-areas/cdiac-appstate.

Global Change Working Group. Memorandum for the Domestic Policy Council, Framework Convention on Climate Change. January 22, 1991.

Global Climate Protection Act of 1987. Pub. L. 100-204, title XI, December 22, 1987. https://www.govinfo.gov/content/pkg/STATUTE-101/pdf/STATUTE-101-Pg1331.pdf.

Gore, Al. *Earth in the Balance: Ecology and the Human Spirit*. Boston: Houghton Mifflin, 1992.

Gore, Al. Letter to The Honorable Ronald W. Reagan. May 21, 1986.

Gore, Al. Remarks at the SE Regional Climate Change Impacts Meeting, Vanderbilt University, Nashville, Tennessee. June 25, 1997.

Gore, Al. The Interplay of Climate Change, Ozone Depletion, and Human Health, excerpts from remarks at the Conference on Human Health and Global Climate Change. September 11, 1995.

Grandoni, Dino. "The Energy 202: Interior Agency Blocks Group of Archaeologists from Attending Scientific Conference." *Washington Post*, May 3, 2018. https://www.washingtonpost.com/news/powerpost/paloma/the-energy-202/2018/05/03/the-energy-202-interior-agency-blocks-group-of-archaeologists-from-attending-scientific-conference/5aea1d9230fb042db57972ac.

Grandoni, Dino, Brady Dennis, and Chris Mooney. "Scott Pruitt Asks Whether Global Warming Is 'Necessarily Is a Bad Thing.'" *Denver Post*, February 7, 2018. https://www.denverpost.com/2018/02/07/scott-pruitt-views-on-climate-change.

Greenhouse Effect and Global Climate Change, Hearing before the Committee on Energy and Natural Resources. 100th Congress. June 23, 1988.

Grid Memo. May 29, 2018. https://www.documentcloud.org/documents/4491203-Grid-Memo.html.

Gustin, Georgina. "Tax Overhaul Preserves Critical Credits for Wind, Solar and Electric Vehicles." *Inside Climate News*, December 22, 2017. https://insideclimatenews.org/news/18122017/tax-bill-vote-renewable-credits-solar-wind-clean-energy-jobs-evs-investment-anwr.

H.Res. 6. Adopting the Rules of the House of Representatives for the One Hundred Sixteenth Congress, and for other purposes. 116th Congress. 2019.

H.Res. 109. Recognizing the Duty of the Federal Government to Create a Green New Deal. 116th Congress. 2019.

"Halliburton Loophole, The." *New York Times*, November 2, 2009. https://www.nytimes.com/2009/11/03/opinion/03tue3.html.

Hammel, Edward F. "Los Alamos Scientific Laboratory Energy-Related History, Research, Managerial Reorganization Proposals, Actions Taken, and Results, 1945–1979." March 1997. http://www.iaea.org/inis/collection/NCLCollectionStore/_Public/28/077/28077313.pdf.

Hand, Mark. "Climate, Environmental Programs Left Mostly Untouched in Budget Deal." *Think Progress*, May 1, 2017. https://archive.thinkprogress.org/climate-environmental-programs-left-mostly-untouched-in-budget-deal-3742f7bad9c5/.

Hansen, James. *Storms of My Grandchildren: The Truth About the Coming Climate Catastrophe and Our Last Chance to Save Humanity*. New York: Bloomsbury, USA, 2009.

Hansen, James, et al. "Target Atmospheric CO_2: Where Should Humanity Aim?" *Open Atmospheric Science Journal* 2 (2008): 217–231.

Harvard Environmental & Energy Law Program. "EPA Mission Tracker." https://eelp.law.harvard.edu/epa-mission-tracker.

Harvard Environmental & Energy Law Program. "Regulatory Rollback Tracker." https://eelp.law.harvard.edu/regulatory-rollback-tracker.

Hennessey, Keith, and Bryan Hannegan. Memorandum to the President through Al Hubbard, Energy Policy—Interim Report. September 30, 2005.

Herrington, John S., DOE Secretary. Letter to the President. March 16, 1987.

Hill, Alan, Chairman, Council on Environmental Quality. Memorandum to Heads of Federal Agencies, Draft of Guidance to Federal Agencies Regarding Consideration of Global Climate Change in Preparation of Environmental Documents. June 21, 1989.

House Select Committee on the Climate Crisis. "Solving the Climate Crisis: The Congressional Action Plan for a Clean Energy Economy and a Healthy, Resilient, and Just America." June 2020. https://climatecrisis.house.gov/sites/climatecrisis.house.gov/files/Climate%20Crisis%20Action%20Plan.pdf.

IEA. *World Energy Investment 2018*. Paris, France: Organisation for Economic Co-operation and Development, 2018. https://www.iea.org/wei2018.

Institute of Medicine, National Academy of Sciences, and National Academy of Engineering. *Policy Implications of Greenhouse Warming*. Washington, DC: National Academy Press, 1991. https://doi.org/10.17226/1794.

Intergovernmental Panel on Climate Change. *Climate Change 2001: The Scientific Basis. Contribution of Working Group I to the Third Assessment Report of the Intergovernmental Panel on*

Climate Change. Cambridge: Cambridge University Press, 2001. https://www.ipcc.ch/site/assets/uploads/2018/03/WGI_TAR_full_report.pdf.

Intergovernmental Panel on Climate Change. *Climate Change 2007: Synthesis Report Summary for Policymakers*. 2007. http://www.ipcc.ch/pdf/assessment-report/ar4/syr/ar4_syr_spm.pdf.

Intergovernmental Panel on Climate Change. *Climate Change 2014 Synthesis Report: Summary for Policymakers*. 2014. https://www.ipcc.ch/pdf/assessment-report/ar5/syr/AR5_SYR_FINAL_SPM.pdf.

Intergovernmental Panel on Climate Change. *Climate Change: The IPCC Scientific Assessment*. Cambridge: Cambridge University Press, 1990. https://archive.ipcc.ch/ipccreports/far/wg_I/ipcc_far_wg_I_full_report.pdf.

Intergovernmental Panel on Climate Change. *IPCC Second Assessment Climate Change 1995: A Report of the Intergovernmental Panel on Climate Change*. 1995. https://archive.ipcc.ch/pdf/climate-changes-1995/ipcc-2nd-assessment/2nd-assessment-en.pdf.

Irfan, Umair. "Trump White House Issues Climate Change Report Undermining Its Own Policy." *Vox*, November 26, 2018. https://www.vox.com/2018/11/26/18112505/national-climate-assessment-2018-trump.

Jacobson, Mark. "Abstracts of 47 Peer-Reviewed Published Journal Articles From 13 Independent Research Groups With 91 Different Authors Supporting the Result That Energy for Electricity, Transportation, Building Heating/Cooling, and/or Industry can be Supplied Reliably with 100% or Near-100% Renewable Energy at Difference Locations Worldwide." December 29, 2019. http://web.stanford.edu/group/efmh/jacobson/Articles/I/CombiningRenew/100PercentPaperAbstracts.pdf.

Johnson, Robert E. Memorandum through Ralph C. Bledsoe for Nancy J. Risque, The Global Climate Protection Act. December 29, 1987.

Johnson, Stephen L., EPA Administrator. Letter to the President. January 31, 2008.

Joyce, Linda A., and Richard Birdsey, technical eds. *The Impact of Climate Change on America's Forests: A Technical Document Supporting the 2000 USDA Forest Service RPA Assessment*. Fort Collins, CO: US Department of Agriculture, Forest Service, Rocky Mountain Research Station, 2000. https://www.fs.fed.us/rm/pubs/rmrs_gtr059.pdf.

Kaplan, Sarah. "Government Scientists Blocked from the Biggest Meeting in Their Field." *Washington Post*, December 22, 2017. https://www.washingtonpost.com/news/speaking-of-science/wp/2017/12/22/government-scientists-blocked-from-the-biggestmeeting-in-their-field.

Kaufman, Alexander C. "Leaked Memo: EPA Shows Workers How to Downplay Climate Change." *HuffPost*, March 28, 2018. https://www.huffpost.com/entry/epa-climate-adaptation_n_5abbb5e3e4b04a59a31387d7.

Kettle, Martin. "Cheney Tells US to Carry on Guzzling." *Guardian*, May 10, 2001. https://www.theguardian.com/world/2001/may/10/dickcheney.martinkettle.

Krosnick, Jon A., et al. "American Opinion on Global Warming: The Impact of the Fall 1997 Debate." *Resources* 133 (1998): 5–9.

Lavelle, Marianne. "Trump's Executive Order: More Fossil Fuels, Regardless of Climate Change." *Inside Climate News*, March 28, 2017. https://insideclimatenews.org/news/28032017/trump-executive-order-climate-change-paris-climate-agreement-clean-power-plan-pruitt.

Levin, Kelly, and Dennis Tirpak. "This Month in Climate Science." World Resources Institute, Washington, DC, August 20, 2020. https://www.wri.org/blog/2020/08/month-climate-science-june-july-2020.

Levitz, Eric. "Trump Thwarts GOP Plot to Pretend His Climate Agenda Isn't Idiotic." *New York*, May 28, 2019. https://nymag.com/intelligencer/2019/05/trump-climate-denier-william-happer-co2-jews-science.html.

Lieberman, Bruce. "What Trump's Proposed NEPA Rollback Could Mean for the Climate." February 20, 2020. https://www.yaleclimateconnections.org/2020/02/what-trumps-proposed-nepa-rollback-could-mean-for-the-climate.

Liptak, Kevin, and Jim Acosta. "Trump on Paris Accord: 'We're Getting Out.'" *CNN*, June 2, 2017. https://www.cnn.com/2017/06/01/politics/trump-paris-climate-decision/index.html.

Loer, David W. Letter to Mr. John Sununu, Chief of Staff. July 30, 1991.

MacDonald, Gordon, et al. *The Long Term Impact of Atmospheric Carbon Dioxide on Climate. Technical Report JSR-78–07*. SRI International, April 1979.

Mahlman, J. D., D. Albritton, and R. T. Watson. State of Scientific Understanding of Climate Change. included in Katie McGinty. Memorandum to the President and the Vice President, Climate Change Action Plan. October 8, 1993.

Majumder, Bianca, Sally Hardin, and Claire Moser. "The Impact of the Coronavirus on the Renewable Energy Industry." *Center for American Progress*, April 15, 2020. https://www.americanprogress.org/issues/green/news/2020/04/15/483219/impact-coronavirus-renewable-energy-industry.

Marshall, Paul, and Heidi Schuttenberg. *A Reef Manager's Guide to Coral Bleaching*. Great Barrier Reef Marine Park Authority, 2006. https://www.coris.noaa.gov/activities/reef_managers_guide/reef_managers_guide.pdf.

Massachusetts v. *EPA*, 549 U.S. 497 (2007).

Maynard, Nancy G. Memorandum to D. Allen Bromley, J. Thomas Ratchford, Environment Update—Week of May 26 and June 3, 1991. June 3, 1991.

McGinty, Katie. Memorandum to the President and the Vice President, Climate Change Action Plan. Washington, DC, October 8, 1993.

McGinty, Katie, Dan Albritton, and Jerry Melillo. Press Briefing, Climate Change Briefing. July 24, 1997.

Melillo, Jerry M., Terese Richmond, and Gary W. Yohe, eds. *Climate Change Impacts in the United States: The Third National Climate Assessment.* US Global Change Research Program. Washington, DC: US Government Printing Office, 2014. https://nca2014.globalchange.gov.

Merrill, M. D., B. M. Sleeter, P. A. Freeman, J. Liu, P. D. Warwick, and B. C. Reed. *Federal Lands Greenhouse Gas Emissions and Sequestration in the United States: Estimates for 2005–14.* US Geological Survey Scientific Investigations Report 2018–5131. 2018. https://doi.org/10.3133/sir20185131.

Michaels, Patrick, and Robert Balling Jr. Letter to The Honorable George H. Bush. February 1, 1991.

Mitchell, George J., et al. Letter to Lee Thomas, Administrator, Environmental Protection Agency. September 12, 1986.

Monroe, Rob. "The History of the Keeling Curve." April 3, 2013. https://sioweb.ucsd.edu/programs/keelingcurve/2013/04/03/the-history-of-the-keeling-curve.

Morgan, Rick. Fax to Mike Toman, Post-2000 Option: CO_2, Cap & Emissions Trading for Electricity, Natural Gas, and Transportation Fuels. October 21, 1994.

Moynihan, Daniel P. Memorandum to John Ehrlichman. September 17, 1969. https://www.nixonlibrary.gov/sites/default/files/virtuallibrary/documents/jul10/56.pdf.

Mulligan, M. "Tackling Climate Change in the United States: The Potential Contribution from Wind Power." Preprint, Conference Paper, National Renewable Energy Laboratory, July 2006.

Narum, David. "A Troublesome Legacy: The Reagan Administration's Conservation and Renewable Energy Policy." *Energy Policy* 20, no.1 (1992): 40–53. http://www.sciencedirect.com/science/article/pii/030142159290146S.

National Academy of Sciences. *Weather and Climate Modification Problems and Prospects, Volume I: Summary and Recommendations.* Final Report of the Panel on Weather and Climate Modification to the Committee on Atmospheric Sciences. Washington, DC: National Academy of Sciences–National Research Council, 1966.

National Aeronautics and Space Administration. Global Mean CO_2 Mixing Ratios. Accessed July 2018. https://data.giss.nasa.gov/modelforce/ghgases/Fig1A.ext.txt.

National Aeronautics and Space Administration, Office of Inspector General. *Investigative Summary Regarding Allegations that NASA Suppressed Climate Change Science and Denied Media Access to Dr. James E. Hansen, a NASA Scientist.* June 2, 2008. https://oig.nasa.gov/docs/OI_STI_Summary.pdf.

National Defense Authorization Act for Fiscal Year 2018. https://www.govinfo.gov/content/pkg/BILLS-115hr2810enr/pdf/BILLS-115hr2810enr.pdf.

National Energy Policy Development Group. *National Energy Policy.* Washington, DC, May 2001. https://www.nrc.gov/docs/ML0428/ML042800056.pdf.

National Geographic. "A Running List of How President Trump Is Changing Environmental Policy." https://www.nationalgeographic.com/news/2017/03/how-trump-is-changing-science-environment.

National Highway Traffic Safety Administration. "Corporate Average Fuel Economy." https://www.nhtsa.gov/laws-regulations/corporate-average-fuel-economy.

National Intelligence Council. *Implications for US National Security of Anticipated Climate Change*. NIC WP 2016–01, September 21, 2016. https://www.dni.gov/files/documents/Newsroom/Reports%20and%20Pubs/Implications_for_US_National_Security_of_Anticipated_Climate_Change.pdf.

National Oceanic and Atmospheric Administration. "Global Climate: A Variable and Vulnerable Natural Resource." October 1, 1986.

National Oceanic and Atmospheric Administration. *Reports to the Nation on Our Changing Planet: The Climate System*. Winter 1991.

National Oceanic and Atmospheric Administration. Trends in Atmospheric Carbon Dioxide, Mauna Loa CO_2 Annual Mean Data. ftp://aftp.cmdl.noaa.gov/products/trends/co2/co2_annmean_mlo.txt.

National Oceanic and Atmospheric Administration Earth System Research Laboratory. Trends in Atmospheric Carbon Dioxide. Accessed July 2018. https://www.esrl.noaa.gov/gmd/ccgg/trends/data.html.

National Oceanic and Atmospheric Administration National Centers for Environmental Information. "State of the Climate: Hurricanes and Tropical Storms—Annual 2005." January 2006. Accessed on July 26, 2016. https://www.ncdc.noaa.gov/sotc/tropical-cyclones/200513.

National Oceanic and Atmospheric Administration, Office of Coast Survey. "Historical Geographic Place Names Removed from NOAA Charts." Edition 3.0, updated August 4, 2014. https://historicalcharts.noaa.gov/pdfs/HistoricalPlacenames_Louisiana.pdf.

National Renewable Energy Laboratory. *Renewable Electricity Futures Study Volume 1: Exploration of High-Penetration Renewable Electricity Futures*. Golden, CO: National Renewable Energy Laboratory, June 2012. https://www.nrel.gov/docs/fy12osti/52409-1.pdf.

National Research Council. *Advancing the Science of Climate Change*. Washington, DC: National Academies Press, 2010. https://doi.org/10.17226/12782.

National Research Council. *Carbon Dioxide and Climate: A Second Assessment*. Washington, DC: National Academy Press, 1982. https://doi.org/10.17226/18524.

National Research Council. *Changing Climate: Report of the Carbon Dioxide Assessment Committee*. Washington, DC: National Academy Press, 1983. https://doi.org/10.17226/18714.

National Research Council. *Climate Change Science: An Analysis of Some Key Questions*. Washington, DC: National Academy Press, 2001. https://doi.org/10.17226/10139.

National Research Council. *Climate Stabilization Targets: Emissions, Concentrations, and Impacts over Decades to Millennia*. Washington, DC: National Academies Press, 2011. https://doi.org/10.17226/12877.

National Research Council. *Energy and Climate: Studies in Geophysics*. Washington, DC: National Academy of Science, 1977. https://doi.org/10.17226/12024.

National Research Council. *Implementing Climate Change and Global Change Research*: A *Review of the Final U.S. Climate Change Science Program Strategic Plan*. Washington, DC: National Academies Press, 2004. https://doi.org/10.17226/10635.

National Science and Technology Council and the Institute of Medicine/National Academy of Sciences. *Conference on Human Health and Global Climate Change: Summary of the Proceedings*. Washington, DC: National Academy Press, 1996.

National Science Foundation, Report on International Geophysical Year, Hearings before the Subcommittee of the Committee on Appropriations. 85th Congress. May 1, 1957.

National Science Foundation. *Weather and Climate Modification: Report of the Special Commission on Weather Modification*. 1965. https://www.nsf.gov/nsb/publications/1965/nsb1265.pdf.

NCAR UCAR. "Stephen Schneider: An Extraordinary Life." July 21, 2010. https://news.ucar.edu/2270/stephen-schneider-extraordinary-life.

Negin, Elliott. "Energy Department Scientists Barred from Attending Nuclear Power Conference." *HuffPost*, August 1, 2017. https://www.huffpost.com/entry/energy-department-scientists-barred-from-attending_b_597f7e2ee4b0cb4fc1c73b8e.

Nitze, William A. "A Failure of Presidential Leadership." In *Negotiating Climate Change: The Inside Story of the Rio Convention*, 187–200. Cambridge: Cambridge University Press, 1994.

Nixon, Richard. Annual Message to the Congress on the State of the Union. January 22, 1970. https://www.presidency.ucsb.edu/node/241063.

Nixon, Richard. Letter to Lee DuBridge, Director, Office of Science and Technology. November 20, 1969.

Oakes, John B. "For Reagan, a Ticking Ecological 'Time Bomb.'" *New York Times*, January 20, 1981. https://www.nytimes.com/1981/01/20/opinion/for-reagan-a-ticking-ecological-time-bomb.html.

Obama, Barack. "Obama's Big Climate Action Plan Announcement." June 25, 2013. https://www.youtube.com/watch?v=TC17DJl6-Ck.

Obama, Barack. "The President's Clean Power Plan." Accessed August 8, 2018, https://obamawhitehouse.archives.gov/node/279886#section-clean-power-plan.

Office of Management and Budget. *A Budget for a Better America: Fiscal Year 2020 Budget of the U.S. Government*. Washington, DC: US Government Publishing Office, 2019. https://www.govinfo.gov/content/pkg/BUDGET-2020-BUD/pdf/BUDGET-2020-BUD.pdf.

Office of Science and Technology. "Responses of the Federal Departments and Agencies to the President's Science Advisory Committee Report, 'Restoring the Quality of Our Environment.'" May 1967.

Office of Science and Technology, Energy Policy Staff. Funding Energy Research and Development. August 26, 1970.

Office of Science and Technology Policy. *Biennial Science and Technology Report to Congress: 1983–1984*. Washington, DC, 1985.

Office of Science and Technology Policy. *Climate Change: State of Knowledge*. 1997.

Office of Science and Technology Policy. Intergovernmental Panel on Climate Change Finalizes Report. February 2, 2007. https://www.presidency.ucsb.edu/node/284041.

Oprysko, Caitlin. "'I Don't Believe It': Trump Dismisses Grim Government Report on Climate Change." *Politico*, November 26, 2018. https://www.politico.com/story/2018/11/26/trump-climate-change-report-1016494.

Ozone Depletion, the Greenhouse Effect, and Climate Change, Hearings before the Subcommittee on Environmental Pollution of the Committee on Environment and Public Works. 99th Congress. June 10 and 11, 1986.

Patterson, Brittany. "Govt. Scientist Blocked from Talking About Climate and Fire." *E&E News*, October 31, 2017. https://www.eenews.net/stories/1060065143.

Peterson, Eugene K. "Carbon Dioxide Affects Global Ecology." *Environmental Science & Technology* 3, no. 11 (1969): 1162–1169.

PEW Charitable Trusts, The. *Driving to 54.5 mpg: A History of Fuel Efficiency in the United States*. September 2012. http://www.pewtrusts.org/-/media/assets/2014/06/02/factsheet-graphic-fuel-effiency-timeline-finalsept-2012.pdf.

PEW Environment Group, The. *History of Fuel Economy*. April 2011. http://www.pewtrusts.org/-/media/assets/2011/04/history-of-fuel-economy-clean-energy-factsheet.pdf.

Pipeline and Hazardous Materials Safety Administration. Pipeline Safety: Gas Pipeline Regulatory Reform, Notice of Proposed Rulemaking. 85 Fed. Reg. 35240. June 9, 2020. https://www.govinfo.gov/content/pkg/FR-2020-06-09/pdf/2020-11843.pdf.

Plumer, Brad. "Trump Orders a Lifeline for Struggling Coal and Nuclear Plants." *New York Times*, June 1, 2018. https://www.nytimes.com/2018/06/01/climate/trump-coal-nuclear-power.html.

Porter, Roger B. Memorandum for the President, The Second World Climate Conference. October 23, 1990.

Presidential Memorandum, Construction of the Keystone XL Pipeline. 82 Fed. Reg. 8663. January 30, 2017. https://www.govinfo.gov/content/pkg/FR-2017-01-30/pdf/2017-02035.pdf.

Press, Frank. Memorandum to the President, Carbon Dioxide Increases. May 5, 1980.

Press, Frank. Memorandum to the President, Release of Fossil CO_2 and the Possibility of a Catastrophic Climate Change. July 7, 1977.

Randol, Randy. Facsimile to Phil Cooney. March 22, 2002. http://www.climatefiles.com/exxonmobil/exxonmobil-memo-2002-from-randy-randol-to-phil-cooney-at-ceq.

Reagan, Ronald. Message to the Congress Transmitting the National Energy Policy Plan. July 17, 1981. https://www.reaganlibrary.gov/research/speeches/71781b.

Reagan, Ronald. Report to Congress: United States Government Activities Related to the Greenhouse Effect. January 26, 1988.

Reuters. "Clean Energy Has Shed Nearly 600,000 U.S. Jobs Due to Pandemic: Report." *Reuters*, May 13, 2020. https://www.reuters.com/article/us-usa-jobs-clean-energy/clean-energy-has-shed-nearly-600000-us-jobs-due-to-pandemic-report-idUSKBN22P2TH.

Revkin, Andrew C. "Editor of Climate Reports Resigns," *New York Times*, June 11, 2005. https://www.nytimes.com/2005/06/11/us/national-briefing-washington-editor-of-climate-reports-resigns.html.

Rhodium Group. "Preliminary US Emissions Estimates for 2018." January 8, 2019. https://rhg.com/research/preliminary-us-emissions-estimates-for-2018.

Rice, Doyle. "Buried? Feds to Release Major Climate Report Day after Thanksgiving." *USA Today*, November 21, 2018. https://www.usatoday.com/story/news/nation/2018/11/21/climate-change-report-released-friday-after-thanksgiving/2080298002.

Rich, Nathaniel. *Losing Earth, A Recent History*. New York: Farrar, Straus and Giroux, 2019.

Richter-Menge, J., et al. *State of the Arctic Report*. National Oceanic and Atmospheric Administration, October 2006.

Roberts, Sam. "Edward E. David Jr., Who Elevated Science Under Nixon, Dies at 92." *New York Times*, February 28, 2017. https://www.nytimes.com/2017/02/28/science/edward-david-dead-science-adviser-to-nixon.html.

Ross, Tom, and Neal Lott. "A Climatology of Recent Extreme Weather and Climate Events." National Oceanic and Atmospheric Administration, National Climatic Data Center. October 2000. https://repository.library.noaa.gov/view/noaa/13825/noaa_13825_DS1.pdf.

Ruckelshaus, William D. Remarks at Organization for Economic Cooperation and Development. June 21, 1984.

Ruckelshaus, William D. The Role of the Private Sector in Environmental Action, World Industry Conference on Environmental Management. November 14, 1984.

S.Res. 59. A Resolution Recognizing the Duty of the Federal Government to Create a Green New Deal. 116th Congress. 2019.

S.Res. 98. A Resolution Expressing the Sense of the Senate Regarding the Conditions for the United States Becoming a Signatory to Any International Agreement on Greenhouse Gas Emissions Under the United Nations Framework Convention on Climate Change. 105th Congress. 1997.

Sabin Center for Climate Change Law. "Climate Deregulation Tracker." https://climate.law.columbia.edu/climate-deregulation-tracker.

Science News. "Trump, Congress Approve Largest U.S. Research Spending Increase in a Decade." *Science*, March 23, 2018. http://www.sciencemag.org/news/2018/03/updated-us-spending-deal-contains-largest-research-spending-increase-decade.

Scripps Institution of Oceanography. "The Keeling Curve." https://scripps.ucsd.edu/programs/keelingcurve.

Selected Questions and Answers on the President's Climate Change Action Plan. In Kathleen McGinty, Letter to Colleague, October 18, 1993.

Shabecoff, Philip. "Global Warming Has Begun, Expert Tells Senate." *New York Times*, June 24, 1988. https://www.nytimes.com/1988/06/24/us/global-warming-has-begun-expert-tells-senate.html.

Shabecoff, Philip. "Scientist Says Budget Office Altered His Testimony." *New York Times*, May 8, 1989. https://www.nytimes.com/1989/05/08/us/scientist-says-budget-office-altered-his-testimony.html.

Shabecoff, Philip. "Scientists Warn U.S. of Carbon Dioxide Peril." *New York Times*, July 10, 1979. https://www.nytimes.com/1979/07/11/archives/scientists-warn-us-of-carbon-dioxide-peril-advice-on-energy.html.

Shabecoff, Philip. "U.S. Calls for Efforts To Combat Global Environmental Problems." *New York Times*, January 15, 1981. https://www.nytimes.com/1981/01/15/world/us-calls-for-efforts-to-combat-global-envioronmental-problems.html.

Shabecoff, Philip. "U.S. Study Warns of Extensive Problems From Carbon Dioxide Pollution." *New York Times*, January 14, 1981. https://www.nytimes.com/1981/01/14/us/us-study-warns-of-extensive-problems-from-carbon-dioxide-pollution.html.

Skibell, Arianna. "Agency Keeps Scientists From Speaking at Watershed Conference." *Greenwire*, October 23, 2017. https://www.eenews.net/greenwire/2017/10/23/stories/1060064343.

Smith, Christopher, et al. "Current Fossil Fuel Infrastructure Does Not Yet Commit Us to 1.5°C Warming." *Nature Communications* 10, no. 1 (2019). https://doi.org/10.1038/s41467-018-07999-w.

Smith, Richard J., Acting Assistant Secretary of State for Oceans and International Environmental and Scientific Affairs. Letter to Richard Hallgren. January 27, 1988.

Smith, Richard J., Acting Assistant Secretary of State for Oceans and International Environmental and Scientific Affairs. Memorandum to Dr. Ralph Bledsoe, White House Domestic Policy Council. January 15, 1988.

Smith, Richard J. Confidential Information Memorandum to The Secretary, United States Department of State, Preparations for an International Conference on the Environment. May 16, 1989.

Solar Energy Industries Association. "Solar Industry Research Data." Accessed August 3, 2018, https://www.seia.org/solar-industry-research-data.

Solar Energy Research Institute. *CO_2 Emissions From Coal-Fired and Solar Electric Power Plants.* May 1990. https://www.nrel.gov/docs/legosti/old/3772.pdf.

Solar Energy Research, Development, and Demonstration Act of 1974. Pub. L. 93–473, 42 USC §5551. 1974.

Sperling, Gene, Katie McGinty, and Daniel Tarullo. Memorandum for the President, Climate Change Scenarios. September 15, 1997.

Staats, Elmer B., Comptroller General. *U.S. Coal Development—Promises, Uncertainties*. EMD-77–43, Report to Congress. September 22, 1977. https://www.gao.gov/products/EMD-77-43.

Stafford, Robert T., et al. Letter to John Gibbons, Executive Director, US Congress Office of Technology Assessment. December 23, 1986.

Statement of Administration Policy: H.R. 6—Energy Policy Act of 2005. June 14, 2005. https://www.presidency.ucsb.edu/node/274355.

Statement of Administration Policy: S. 3036—Lieberman–Warner Climate Security Act. June 2, 2008. https://www.presidency.ucsb.edu/node/277902.

Stone, Peter. "'Swampy Symbiosis': Fossil Fuel Industry Has More Clout Than Ever Under Trump." *Guardian*, September 27, 2019. https://www.theguardian.com/environment/2019/sep/27/fossil-fuel-industry-clout-trump-era.

Sununu, John. Letter to David W. Loer. August 20, 1991.

Swanson, Ana, and Brad Plumer. "Trump Slaps Steep Tariffs on Foreign Washing Machines and Solar Products." *New York Times*, January 22, 2018. https://www.nytimes.com/2018/01/22/business/trump-tariffs-washing-machines-solar-panels.html.

Talking Points for Governor Christine Todd Whitman, Administrator, United States Environmental Protection Agency at the G8 Environmental Ministerial Meeting Working Session on Climate Change, Trieste, Italy. March 3, 2001.

Talley, Ian, and Siobhan Hughes. "White House Blocks EPA Emissions Draft." *The Wall Street Journal*, June 30, 2008. https://www.wsj.com/articles/SB121478564162114625.

Tenpas, Kathryn Dunn. "Vacancies, Acting Officials and the Waning Role of the U.S. Senate." *The Brookings Institute*, September 24, 2020. https://www.brookings.edu/blog/fixgov/2020/09/24/vacancies-acting-officials-and-the-waning-role-of-the-u-s-senate.

"The Trump Administration Is Reversing Nearly 100 Environmental Rules. Here's the Full List." *New York Times*, Last updated October 15, 2020. https://www.nytimes.com/interactive/2020/climate/trump-environment-rollbacks.html.

Tollefson, Jeff. "U.S. Government Disbands Climate-Science Advisory Committee." *Scientific American*, April 21, 2017. https://www.scientificamerican.com/article/u-s-government-disbands-climate-science-advisory-committee.

"Tracking How Many Key Positions Trump Has Filled So Far." *Washington Post*. https://www.washingtonpost.com/graphics/politics/trump-administration-appointee-tracker/database.

Union of Concerned Scientists. "New UCS Report Tallies Attacks on Science in Trump Era Harming Public Health." January 28, 2019. https://www.ucsusa.org/about/news/ucs-report-tallies-attacks-science.

United Nations Framework Convention on Climate Change. Report of the Conference of the Parties on Its Fifteenth Session, Held in Copenhagen from 7 to 19 December 2009. https://unfccc.int/resource/docs/2009/cop15/eng/11a01.pdf.

United Nations. *Kyoto Protocol to the United Nations Framework Convention on Climate Change.* 1998. https://unfccc.int/resource/docs/convkp/kpeng.pdf.

United Nations. *Our Common Future: Report of the World Commission on Environment and Development.* 1987. https://sustainabledevelopment.un.org/content/documents/5987our-common-future.pdf.

United Nations. *Paris Agreement.* 2015. https://unfccc.int/sites/default/files/english_paris_agreement.pdf.

United Nations. *United Nations Framework Convention on Climate Change.* 1992. https://unfccc.int/resource/docs/convkp/conveng.pdf.

US Climate Change Science Program. *Analyses of the Effects of Global Change on Human Health and Welfare and Human Systems.* A Report by the U.S. Climate Change Science Program and the Subcommittee on Global Change Research. Washington, DC, September 2008. https://downloads.globalchange.gov/sap/sap4-6/sap4-6-final-report-all.pdf.

US Climate Change Science Program. *Our Changing Planet: The U.S. Climate Change Science Program for Fiscal Year 2009.* A Report by the Climate Change Science Program and the Subcommittee on Global Change Research. Washington, DC, July 2008. https://downloads.globalchange.gov/ocp/ocp2009/ocp2009.pdf.

US Climate Change Science Program. *Scenarios of Greenhouse Gas Emissions and Atmospheric Concentrations.* Sub-Report 2.1A of Synthesis and Assessment Product 2.1 by the US Climate Change Science Program and the Subcommittee on Global Change Research. Washington, DC, 2007. https://downloads.globalchange.gov/sap/sap2-1a/sap2-1a-final-all.pdf.

US Congress, Office of Technology Assessment. *Changing by Degrees: Steps to Reduce Greenhouse Gases, OTA-O-482.* Washington, DC: US Government Printing Office, February 1991.

US Congress, Office of Technology Assessment. *Preparing for an Uncertain Climate: Volume 1, OTA-0-567.* Washington DC: US Government Printing Office, October 1993.

US Department of Commerce. *The Automobile and Air Pollution: A Program for Progress.* Report of the Panel on Electrically Powered Vehicles. Washington, DC, 1967.

US Department of Commerce et al. *World Weather Program: Plan for Fiscal Year 1971.* April 1970.

US Department of Defense. *2014 Climate Change Adaptation Roadmap.* 2014. https://www.acq.osd.mil/EIE/Downloads/CCARprint_wForward_e.pdf.

US Department of Energy. *Carbon Dioxide Effects Research and Assessment Program: Environmental and Societal Consequences of a Possible CO_2-Induced Climate Change.* Vol. II, Part I, Response of the West Antarctic Ice Sheet to CO_2-Induced Climatic Warming. Washington, DC, April 1982.

US Department of Energy. *Carbon Dioxide Effects Research and Assessment Program: Proceedings of the Workshop on First Detection of Carbon Dioxide Effects.* Harpers Ferry, West Virginia, June 8–10, 1981. Washington, DC, May 1982. http://hdl.handle.net/2027/uc1.31822016268534.

US Department of Energy. *Carbon Dioxide Effects Research and Assessment Program: Workshop on the Global Effects of Carbon Dioxide from Fossil Fuels*. May 1979. https://www.osti.gov/biblio/6385084.

US Department of Energy. "Chapter I: Transforming the Nation's Electricity System: The Second Installation of the Quadrennial Energy Review." In *Transforming the Nation's Electricity System*. January 2017. https://www.energy.gov/sites/prod/files/2017/02/f34/Chapter%20I--Transforming%20the%20Nation%27s%20Electricity%20System.pdf.

US Department of Energy. "DOE Announces Intent to Provide $122M to Establish Coal Products Innovation Centers." June 26, 2020. https://www.energy.gov/articles/doe-announces-intent-provide-122m-establish-coal-products-innovation-centers.

US Department of Energy. *Million Solar Roofs: Become One in a Million*. Washington, DC, 2003. https://www.nrel.gov/docs/fy04osti/34009.pdf.

US Department of Energy. *National Energy Strategy: Powerful Ideas for America*. Washington, DC, February 1991.

US Department of Energy. "October 24, 1992: Energy Policy Act of 1992." https://www.energy.gov/management/october-24-1992-energy-policy-act-1992.

US Department of Energy. *Strategic Plan*. May 2011. https://www.energy.gov/sites/prod/files/2011_DOE_Strategic_Plan_.pdf.

US Department of Energy. *Summary of the Carbon Dioxide Effects Research and Assessment Program*. July 1980. https://doi.org/10.2172/5102316.

US Department of Energy. *The Economics of Long-Term Global Climate Change: A Preliminary Assessment*. Report of an Interagency Task Force, Washington, DC, September 1990. https://doi.org/10.2172/6487569.

US Department of Energy, John R. Trabalka, ed., *Atmospheric Carbon Dioxide and the Global Carbon Cycle*. Washington, DC, December 1985. https://doi.org/10.2172/6048470.

US Department of Energy, Michael MacCracken and Frederick M. Luther, eds. *Projecting the Climatic Effects of Increasing Carbon Dioxide*. Washington, DC, December 1985.

US Department of Energy, Michael P. Farrell, ed. *Master Index for the Carbon Dioxide Research State-of-the-Art Report Series*. Washington, DC, March 1987. https://doi.org/10.2172/6176904.

US Department of Energy, Office of Fossil Energy. *Environmental Benefits of Advanced Oil and Gas Exploration and Production Technology*. Washington, DC, October 1999. https://doi.org/10.2172/771125.

US Department of Energy, Office of Fossil Energy. Extending Natural Gas Export Authorizations to Non-Free Trade Agreement Countries Through the Year 2050. 85 Fed. Reg. 52237. August 25, 2020. https://www.energy.gov/sites/prod/files/2020/09/f78/2020-16836_FE_Policy%20Statement%20Year%202050.pdf.

US Department of State. "PRD-12/Global Climate Change Policy Decision Paper." https://nsarchive2.gwu.edu//dc.html?doc=4114691-01-PRD-12-Global-Climate-Change-Policy-Decision.

US Department of State. Remarks by the Honorable James A. Baker III, Secretary of State before the Response Strategies Working Group, Intergovernmental Panel on Climate Change. January 30, 1989.

US Department of State. *Second Biennial Report of the United States Under the United Nations Framework Convention on Climate Change.* 2016. https://unfccc.int/files/national_reports/biennial_reports_and_iar/submitted_biennial_reports/application/pdf/2016_second_biennial_report_of_the_united_states_.pdf.

US Department of State. *U.S. Climate Action Report 2002: Third National Communication of the United States of America Under the United Nations Framework Convention on Climate Change.* Washington, DC, May 2002. https://unfccc.int/resource/docs/natc/usnc3.pdf.

US Department of the Interior. *Climatic Change in the National Parks, Wildlife Refuges and Other Department of Interior Lands in the United States.* May 1997.

US Department of the Interior. "Interior Announces Date for Largest Oil and Gas Lease Sale in U.S. History." February 16, 2018. https://www.doi.gov/pressreleases/interior-announces-date-largest-oil-and-gas-lease-sale-us-history.

US Department of the Interior. Sec. Order No. 3350. America-First Offshore Energy Strategy. May 1, 2017. https://www.doi.gov/sites/doi.gov/files/press-release/secretarial-order-3350.pdf.

US Department of the Interior. "Secretary Bernhardt Signs Decision to Implement the Coastal Plain Oil and Gas Leasing Program in Alaska." August 17, 2020. https://www.doi.gov/pressreleases/secretary-bernhardt-signs-decision-implement-coastal-plain-oil-and-gas-leasing-program.

US Department of the Interior. "Secretary Zinke Announces Largest Oil & Gas Lease Sale in U.S. History." October 24, 2017. https://www.doi.gov/pressreleases/secretary-zinke-announces-largest-oil-gas-lease-sale-us-history.

US Department of the Interior. "Secretary Zinke Announces Plan For Unleashing America's Offshore Oil and Gas Potential." January 4, 2018. https://www.doi.gov/pressreleases/secretary-zinke-announces-plan-unleashing-americas-offshore-oil-and-gas-potential.

US Department of the Interior, Bureau of Ocean Energy Management. *2019–2024 National Outer Continental Shelf Oil and Gas Leasing, Draft Proposed Program.* January 2018. https://www.boem.gov/sites/default/files/oil-and-gas-energy-program/Leasing/Five-Year-Program/2019-2024/DPP/NP-Draft-Proposed-Program-2019-2024.pdf.

US Department of Transportation. *Summary of Fuel Economy Performance.* Washington, DC, December 15, 2014. https://www.nhtsa.gov/sites/nhtsa.dot.gov/files/performance-summary-report-12152014-v2.pdf.

US Energy Information Administration. "Energy Use & Related Statistics: Carbon Dioxide Emissions." https://www.eia.gov.

US Energy Information Administration. *Monthly Energy Review April 2020.* 2020. https://www.eia.gov/totalenergy/data/monthly/archive/00352004.pdf.

US Energy Information Administration. *Monthly Energy Review August 2020*. 2020. https://www.eia.gov/totalenergy/data/monthly/archive/00352008.pdf.

US Energy Information Administration. *Monthly Energy Review March 2018*. 2018. https://www.eia.gov/totalenergy/data/monthly/archive/00351803.pdf.

US Energy Information Administration. *Monthly Energy Review May 2018*. 2018. https://www.eia.gov/totalenergy/data/monthly/archive/00351805.pdf.

US Energy Information Administration. *Sales of Fossil Fuels Produced from Federal and Indian Lands, FY 2003 through FY 2014*. Washington, DC, July 2015. https://www.eia.gov/analysis/requests/federallands/pdf/eia-federallandsales.pdf.

US Environmental Protection Agency. *America's Children and the Environment*, 3rd ed. Washington, DC, January 2013. https://www.epa.gov/sites/production/files/2015-06/documents/ace3_2013.pdf.

US Environmental Protection Agency. *Can We Delay a Greenhouse Warming?* Washington, DC, September 1983.

US Environmental Protection Agency. "Climate Change." https://19january2017snapshot.epa.gov/climatechange_.html.

US Environmental Protection Agency. "Climate Change Indicators: Climate Forcing." https://www.epa.gov/climate-indicators/climate-change-indicators-climate-forcing.

US Environmental Protection Agency. *Climate Change Indicators in the United States, 2012*. 2nd ed. Washington, DC, December 2012. https://www.epa.gov/sites/production/files/2016-08/documents/climateindicators-full-2012.pdf.

US Environmental Protection Agency. *Climate Change Indicators in the United States, 2014*. 3rd ed. Washington, DC, May 2014. https://www.epa.gov/sites/production/files/2016-07/documents/climateindicators-full-2014.pdf.

US Environmental Protection Agency. *Effects of CO_2 and Climate Change on Forest Trees*. Environmental Research Laboratory—Corvallis, April 1993.

US Environmental Protection Agency. Endangerment and Cause or Contribute Findings for Greenhouse Gases Under Section 202(a) of the Clean Air Act, Final Rule. 74 Fed. Reg. 66496. December 15, 2009. https://www.govinfo.gov/content/pkg/FR-2009-12-15/pdf/E9-29537.pdf.

US Environmental Protection Agency. *Environmental Health Threats to Children*. Washington, DC, September 1996.

US Environmental Protection Agency. "EPA Issues Final Policy and Technical Amendments to the New Source Performance Standards for the Oil and Natural Gas Industry." https://www.epa.gov/controlling-air-pollution-oil-and-natural-gas-industry/epa-issues-final-policy-and-technical.

US Environmental Protection Agency. "EPA Takes Another Step to Advance President Trump's America First Strategy, Proposes Repeal of 'Clean Power Plan.'" October 10, 2017. https://archive.epa.gov/epa/newsreleases/epa-takes-another-step-advance-president-trumps-america-first-strategy-proposes-repeal.html.

US Environmental Protection Agency. "Final FY 2018–2019 Office of Land and Emergency Management National Program Manager Guidance." 540B17001. Sept. 29, 2017.

US Environmental Protection Agency. "Final FY 2020–2021 Office of Land and Emergency Management National Program Manager Guidance." 500B19002. June 7, 2019.

US Environmental Protection Agency. *Inventory of U.S. Greenhouse Gas Emissions and Sinks: 1990–2016*. 2018. https://www.epa.gov/sites/production/files/2018-01/documents/2018_complete_report.pdf.

US Environmental Protection Agency. Oil and Natural Gas Sector: Emission Standards for New, Reconstructed, and Modified Sources: Stay of Certain Requirements: Proposed Rule. 82 Fed. Reg. 27645. June 16, 2017. https://www.govinfo.gov/content/pkg/FR-2017-06-16/pdf/2017-12698.pdf.

US Environmental Protection Agency. Oil and Natural Gas Sector: Emission Standards for New, Reconstructed, and Modified Sources Review: Final Rule. 85 Fed. Reg. 57018. September 14, 2020. https://www.govinfo.gov/content/pkg/FR-2020-09-14/pdf/2020-18114.pdf.

US Environmental Protection Agency. *Policy Options for Stabilizing Global Climate: Report to Congress*. December 1990.

US Environmental Protection Agency. *Projecting Future Sea Level Rise: Methodology, Estimates to the Year 2100, and Research Needs*. Washington, DC, October 1983.

US Environmental Protection Agency. "Regulations for Greenhouse Gas Emissions from Passenger Cars and Trucks." https://www.epa.gov/regulations-emissions-vehicles-and-engines/regulations-greenhouse-gas-emissions-passenger-cars-and.

US Environmental Protection Agency. Regulating Greenhouse Gas Emissions Under the Clean Air Act, Advanced Notice of Proposed Rulemaking. 73 Fed. Reg. 44354. July 30, 2008. https://www.govinfo.gov/content/pkg/FR-2008-07-30/pdf/E8-16432.pdf.

US Environmental Protection Agency. Regulatory Determination for Oil and Gas and Geothermal Exploration, Development and Production Wastes. 53 Fed. Reg. 25446. July 6, 1988. https://archive.epa.gov/epawaste/nonhaz/industrial/special/web/pdf/og88wp.pdf.

US Environmental Protection Agency. Repeal of the Clean Power Plan; Emission Guidelines for Greenhouse Gas Emissions From Existing Electric Utility Generating Units; Revisions to Emission Guidelines Implementing Regulations: Final Rule. 84 Fed. Reg. 32520. July 8, 2019. https://www.federalregister.gov/documents/2019/07/08/2019-13507/repeal-of-the-clean-power-plan-emissionguidelines-for-greenhouse-gas-emissions-from-existing.

US Environmental Protection Agency. *The Potential Effects of Global Climate Change on the United States*. Washington, DC, December 1989.

US Environmental Protection Agency. "The Safer Affordable Fuel Efficient (SAFE) Vehicles Proposed Rule for Model Years 2021–2026." https://www.epa.gov/regulations-emissions-vehicles-and-engines/safer-and-affordable-fuel-efficient-vehicles-proposed.

US Environmental Protection Agency. *Working Together: FY 2018–2022 U.S. EPA Strategic Plan.* EPA-190-R-18–003. Washington, DC, February 2018. https://www.epa.gov/sites/production/files/2018-02/documents/fy-2018-2022-epa-strategic-plan.pdf.

US Environmental Protection Agency and National Highway Traffic Safety Administration. 2017 and Later Model Year Light-Duty Vehicle Greenhouse Gas Emissions and Corporate Average Fuel Economy Standards: Final Rule. 77 Fed. Reg. 62624. October 15, 2012. https://www.govinfo.gov/content/pkg/FR-2012-10-15/pdf/2012-21972.pdf.

US Environmental Protection Agency and National Highway Traffic Safety Administration. Light-Duty Vehicle Greenhouse Gas Emission Standards and Corporate Average Fuel Economy Standards; Final Rule. 75 Fed. Reg. 25325. May 7, 2010. https://www.govinfo.gov/content/pkg/FR-2010-05-07/pdf/2010-8159.pdf.

US General Accounting Office. *Energy Conservation: DOE's Efforts to Promote Energy Conservation and Efficiency.* Washington, DC, April 1992. http://archive.gao.gov/d32t10/146799.pdf.

US General Accounting Office. *Energy Policy: Options to Reduce Environmental and Other Costs of Gasoline Consumption.* Washington, DC, September 1992. https://www.gao.gov/assets/160/152332.pdf.

US General Accounting Office. *Global Warming: Administration Approach Cautious Pending Validation of Threat.* Washington, DC, January 1990. https://www.gao.gov/assets/150/148577.pdf.

US General Accounting Office. *Global Warming: Administration's Proposal in Support of the Kyoto Protocol.* Statement of Victor S. Rezendes, Testimony Before the Committee on Energy and Natural Resources. June 4, 1998. https://www.govinfo.gov/content/pkg/GAOREPORTS-T-RCED-98-219/pdf/GAOREPORTS-T-RCED-98-219.pdf.

US General Accounting Office. *Global Warming: Emission Reductions Possible as Scientific Uncertainties are Resolved.* Washington, DC, September 1990. https://www.gao.gov/assets/150/149882.pdf.

US General Accounting Office. *Greenhouse Effect: DOE's Programs and Activities Relevant to the Global Warming Phenomenon.* Washington, DC, March 1990. https://www.gao.gov/assets/80/77774.pdf.

US Geological Survey. "Coastal Wetlands and Global Change: Overview." June 1997.

US Geological Survey. "Science-Based Strategies for Sustaining Coral Ecosystems." Fact Sheet 2009–3089. September 2009. https://pubs.usgs.gov/fs/2009/3089/pdf/brewercoralfs3.pdf.

US Global Change Research Program. *Climate Science Special Report: Fourth National Climate Assessment, Volume I.* Washington, DC, 2017. https://science2017.globalchange.gov.

US Global Change Research Program. *Global Climate Change Impacts in the United States.* New York: Cambridge University Press, 2009. https://downloads.globalchange.gov/usimpacts/pdfs/climate-impacts-report.pdf.

US Global Change Research Program. *Impacts, Risks, and Adaptation in the United States: Fourth National Climate Assessment, Volume II.* Washington, DC, 2018. https://nca2018.globalchange.gov.

US Global Change Research Program. *The Impacts of Climate Change on Human Health in the United States: A Scientific Assessment.* Washington, DC, 2016. https://health2016.globalchange.gov.

US Government Accountability Office. *Advanced Energy Technologies: Budget Trends and Challenges for DOE's Energy R&D Program.* GAO-08-556T. Statement of Mark E. Gaffigan, Testimony Before the Subcommittee on Energy and Environment, Committee on Science and Technology, House of Representatives. March 5, 2008. https://www.gao.gov/new.items/d08556t.pdf.

US Government Accountability Office. *Advanced Energy Technologies: Key Challenges to Their Development and Deployment.* GAO-07–550T. Statement of Jim Wells, Testimony Before the Subcommittee on Energy and Water Development, Committee on Appropriations, House of Representatives. February 28, 2007. https://www.gao.gov/assets/120/115602.pdf.

US Government Accountability Office. *Climate Information: A National System Could Help Federal, State, Local, and Private Sector Decision Makers Use Climate Information.* GAO-16–37. Washington, DC, November 2015.

US Government Accountability Office. *Department of Energy: Key Challenges Remain for Developing and Deploying Advanced Energy Technologies to Meet Future Needs.* GAO-07-106. December 2006. https://www.gao.gov/new.items/d07106.pdf.

US Government Accountability Office. *High-Risk Series: Substantial Efforts Needed to Achieve Greater Progress on High-Risk Areas.* GAO-19–157SP. Washington, DC, March 6, 2019. https://www.gao.gov/assets/700/697245.pdf.

US Government Accountability Office. *Renewable Energy: Wind Power's Contribution to Electric Power Generation and Impact of Farms and Rural Communities.* GAO-04–756. Washington, DC, September 2004. https://www.gao.gov/new.items/d04756.pdf.

US Government Accountability Office. *Social Cost of Carbon: Identifying a Federal Entity to Address the National Academies' Recommendations Could Strengthen Regulatory Analysis.* GAO-20-254. Washington, DC, June 2020. https://www.gao.gov/assets/710/707871.pdf.

Visser, Nick. "Interior Department Scrubs Climate Change From Agency Website. Again." *HuffPost,* June 13, 2017. https://www.huffpost.com/entry/interior-department-climate-change_n_593f8bcae4b0b13f2c6d8a9a.

Washington Post Editorial Board. "Pruitt and Perry Continue to Play Down Climate Change." *Washington Post,* January 21, 2017. https://www.washingtonpost.com/opinions/pruitt-and-perry-continue-to-play-down-climate-change/2017/01/21/c891c61c-de97-11e6-ad42-f3375f271c9c_story.html.

Watson, Harlan. "U.S. Climate Change Policy." Presentation at US-Germany Bilateral Meeting, Berlin, Germany, August 12, 2005.

Weisskopf, Michael. "Bush Was Aloof in Warming Debate." *Washington Post,* October 31, 1992. https://www.washingtonpost.com/archive/politics/1992/10/31/bush-was-aloof-in-warming-debate/f14bea92-884c-401b-9870-5bef72960806.

Wheeler, Andrew. Memorandum to Assistant Administrators, Increasing Consistency and Transparency in Considering Benefits and Costs in the Rulemaking Process. May 13, 2019. https://www.epa.gov/environmental-economics/administrator-wheeler-memorandum-increasing-consistency-and-transparency.

White House, The. "Energy & Environment." Accessed August 6, 2018. https://www.whitehouse.gov/issues/energy-environment.

White House, The. "Global Climate Change: An East Room Roundtable." July 24, 1997.

White House, The. *National Security Strategy of the United States of America.* December 2017.

White House, The. "President Donald J. Trump Is Unleashing American Energy Dominance." May 14, 2019. https://trumpwhitehouse.archives.gov/briefings-statements/president-donald-j-trump-unleashing-american-energy-dominance/.

White House, The. *Restoring the Quality of Our Environment: Report of the Environmental Pollution Panel, President's Science Advisory Committee.* Washington, DC, November 1965.

White House, The. *United States Mid-Century Strategy for Deep Decarbonization.* Washington, DC, November 2016. https://obamawhitehouse.archives.gov/sites/default/files/docs/mid_century_strategy_report-final.pdf.

"White House and the Greenhouse, The." *New York Times,* May 9, 1989. https://www.nytimes.com/1989/05/09/opinion/the-white-house-and-the-greenhouse.html.

White House, The. Office of the Press Secretary. Fact Sheet: President Obama's 21st Century Clean Transportation System. February 4, 2016. https://obamawhitehouse.archives.gov/the-press-office/2016/02/04/fact-sheet-president-obamas-21st-century-clean-transportation-system.

White House, The. Office of the Press Secretary. Fact Sheet: U.S. Reports Its 2025 Emissions Target to the UNFCCC. March 31, 2015. https://obamawhitehouse.archives.gov/the-press-office/2015/03/31/fact-sheet-us-reports-its-2025-emissions-target-unfccc.

White House, The. Office of the Press Secretary. Inaugural Address by President Barack Obama. January 21, 2013. https://obamawhitehouse.archives.gov/the-press-office/2013/01/21/inaugural-address-president-barack-obama.

White House, The. Office of the Press Secretary. Opening Remarks by the President and the Vice President at Discussion on Climate Change. July 24, 1997.

White House, The. Office of the Press Secretary. President Bush Discusses Global Climate Change, The Rose Garden. June 11, 2001.

White House, The. Office of the Press Secretary. President Discusses Energy Policy, Franklin County Veterans Memorial, Columbus, Ohio. March 9, 2005. https://georgewbush-whitehouse.archives.gov/news/releases/2005/03/20050309-5.html.

White House, The. Office of the Press Secretary. Remarks by the President at the White House Conference on Climate Change, Georgetown University. October 6, 1997.

White House, The. Office of the Press Secretary. Remarks by the President on Global Climate Change, National Geographic Society, Washington, DC. October 22, 1997.

White House, The. Office of the Press Secretary. Remarks by the President to Business Roundtable, Washington, DC. June 12, 1997.

White House, The. Office of the Press Secretary. Remarks by the President in the State of the Union Address. February 12, 2013. https://obamawhitehouse.archives.gov/the-press-office/2013/02/12/remarks-president-state-union-address.

White House, The. Office of the Press Secretary. Remarks by the President on American-Made Energy. March 22, 2012. https://obamawhitehouse.archives.gov/the-press-office/2012/03/22/remarks-president-american-made-energy.

White House, The. Office of the Press Secretary. Remarks by the President on Climate Change, Georgetown University, Washington, DC. June 25, 2013. https://obamawhitehouse.archives.gov/the-press-office/2013/06/25/remarks-president-climate-change.

White House, The. Office of the Press Secretary. Remarks by the President in State of the Union Address. January 24, 2012. https://obamawhitehouse.archives.gov/the-press-office/2012/01/24/remarks-president-state-union-address.

Whitman, Christine Todd, Administrator of the US Environmental Protection Agency. Remarks at the Business Council, Washington, DC. February 22, 2001.

Wirth, Timothy E., Undersecretary for Global Affairs on Behalf of the United States of America. Statement at the Second Conference of the Parties, Framework Convention on Climate Change. July 17, 1996.

Wirth, Timothy E., Undersecretary of State. Remarks at the First Conference of the Parties to the Framework Convention on Climate Change. April 5, 1995.

Woodwell, George M., Gordon J. MacDonald, Roger Revelle, and C. David Keeling. *The Carbon Dioxide Problem: Implications for Policy in the Management of Energy and Other Resources.* A Report to the Council on Environmental Quality. July 1979, reprinted 2008. https://static01.nyt.com/packages/pdf/science/woodwellreport.pdf.

World Climate Programme. *Report of the International Conference on the Assessment of the Role of Carbon Dioxide and of Other Greenhouse Gases in Climate Variations and Associated Impacts.* Villach, Austria, 9–15 October 1985. 1986. https://library.wmo.int/doc_num.php?explnum_id=8512.

World Meteorological Organization. "Conference Proceedings: The Changing Atmosphere: Implications for Global Security, Toronto, Canada, 27–30 June 1988." 1988. https://wedocs.unep.org/handle/20.500.11822/29980.

Index

Page numbers followed by an "f" indicate figures.